REVOLT AMONG THE SHARECROPPERS

SOUTHERN TENANT FARMERS UNION

HOWARD KESTER

Revolt Among the Sharecroppers

With an Introduction by Alex Lichtenstein

THE UNIVERSITY OF TENNESSEE PRESS · KNOXVILLE

Paper: 1st printing, 1997; 2nd printing, 1999.

The paper used in this book meets the minimum requirements of ANSI/NISO Z39.48-1992 (R 1997) (Permanence of Paper). The binding materials have been chosen for strength and durability. Printed on recycled paper.

Cover photograph and frontispiece: This 1937 photograph of Mississippi Delta cotton workers, taken by Margaret Bourke-White for the Farm Security Administration (FSA), is typical of the work produced by documentary photographers during the 1930s. Kester and other leaders of the Southern Tenant Farmers' Union expressed skepticism about the limitations of the FSA's rural rehabilitation program. Nevertheless, nothing did more to unveil rural poverty than the unforgettable images taken by Bourke-White, Dorothea Lange, Arthur Rothstein, Walker Evans, and other FSA photographers. Southern Historical Collection, Library of the University of North Carolina at Chapel Hill.

Also on cover: Southern Tenant Farmers' Union (STFU) seal. Like the civil rights movement of a later generation, the STFU drew on the talents and commitment of socially active young people from privileged backgrounds. This logo, designed in 1937 by Janet Frazier, an STFU student volunteer from Bennington College, appeared on the union's stationery, membership booklets, and official documents. Southern Historical Collection, Library of the University of North Carolina at Chapel Hill.

Division page: The original cover of *Revolt among the Sharecroppers* as it appeared upon publication as a pamphlet in March 1936. For the cover, Kester created a montage of many of the striking photographs he took while working with the STFU throughout 1935. He donated all of the proceeds of the pamphlet's sales to the union. STFU Papers, Southern Historical Collection, Library of the University of North Carolina at Chapel Hill.

Library of Congress Cataloging-in-Publication Data

Kester, Howard, 1904–1977.
Revolt among the sharecroppers / Howard Kester; with an introduction by Alex Lichtenstein. — 1st ed.
p. cm.
Originally published: New York: Covici, Friede, [c1936]
ISBN 0-87049-975-0 (pbk.: alk. paper)
1. Tenant farmers—United States—History. 2. Sharecroppers— Arkansas. 3. Cotton growing—Arkansas. 4. Southern Tenant Farmers' Union. 5. United States. Agricultural Adjustment Administration. I. Title.
HD1511.U5K4 1997
333.33'5563'09767—dc20 96-35712

CONTENTS

ILLUSTRATIONS

PREFACE

In 1991 I taught for the first time a course on the history of the 1930s. In order to help my students grasp the significance of one of the most fascinating rural social movements in American history, I asked them to read Howard Kester's *Revolt among the Sharecroppers*. They must have acquired the last available copies of this account of the formation of the interracial Southern Tenant Farmers' Union (STFU), originally published in 1936 and reprinted in 1969 by Arno Press in the series "The American Negro, His History and His Literature." When I ordered the book the following year to use in my African American history class it had become unavailable.

Disappointed, I began a lonely campaign to have this remarkable text reissued, mostly so I myself could continue to introduce it to my students and benefit from its utility as a teaching tool. I soon realized, however, that other faculty at the college level might find *Revolt among the Sharecroppers* as useful in the classroom as I did. As a primary document made available to students, I think Kester's book has many strengths. First, it is short. Second, it tells a dramatic and compelling story from the point of view of a participant in one of the most exciting social movements in the pre–civil rights South. Finally, in an accessible fashion, it helps draw out quite a few key features of modern southern history, especially the impact of the Great Depression and the New Deal on rural life and labor.

I hope *Revolt among the Sharecroppers* finds its deserved place on the syllabus in surveys of twentieth-century America and in more specific topical courses. For a class on the history of the postbellum South it offers an excellent, if brief, description of the origins, development, and demise of the sharecropping system. This is a saga that had a tremendous impact on the course of modern American history, urban as well as rural. Moreover there are few better accounts of the necessity of—and barriers to—workers' interracial solidarity under the plantation system, and by implication in the great wave of industrial unionism that swept the nation's factories and mines during the 1930s. In African American history the book evokes the realities of rural black life in the early twentieth century and suggests how the 1930s opened up new possibilities for southern blacks to challenge their condition.

As a commentator on the New Deal, Kester gives a concise critique of the Agricultural Adjustment Administration's deleterious effects on tenants and sharecroppers, a tragedy with which all students of modern America should be made familiar. The text itself, and the story of its protagonists, is a classic example of Depression-era social activism by those on the socialist and religious left who championed a far more radical vision of social change than did the liberals who shaped the New Deal. Finally, labor historians desperately need an available work on the Southern Tenant Farmers' Union and its efforts at interracial organizing among agricultural workers, a group all too often ignored in courses on labor history. I imagine that for many of the same reasons this primary document would be useful in courses on the history of the American Left as well, if anyone still teaches such a class.

In my introductory remarks I offer students and faculty some guideposts to these issues, along with the historical backdrop for the sharecropping system in the Arkansas Delta, Kester's radicalism, and the STFU's program and limitations. Of course, as an organization the STFU was unsuccessful, failing to live up to some of Kester's loftier ideals. I try to suggest some of the reasons for this, but I also point to the remarkable rebirth during the 1960s of the union's most enduring contributions to the history of American social movements.

I hasten to add that, while I envision *Revolt among the Sharecroppers* as a potential addition to college syllabi, this lost classic of American radicalism merits republication on its own terms. Written during the nation's worst economic crisis, it should still appeal to all Americans who object to our current widening gap between rich and poor and yearn for a more just political economy. An unusually prophetic plea for interracial brotherhood during the reign of Jim Crow, it still speaks to those who await full racial equality. Confident in the ability of ordinary people to transform the conditions of their own lives, it reminds us that great movements for social change can arise in the least expected places. In short, Kester's impassioned call for social justice carries powerful reverberations down to our own time.

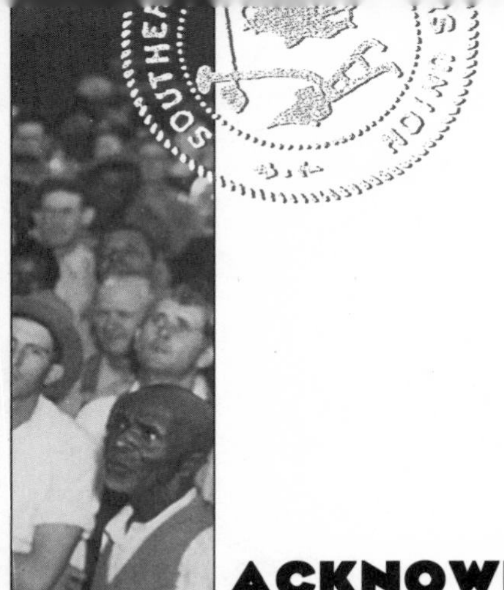

ACKNOWLEDGMENTS

Several folks deserve acknowledgment for helping to make this new edition a reality. Meredith Morris-Babb at the University of Tennessee Press greeted the idea with enthusiasm from the start. Edward Ayers, Michael Honey, Robert Martin, and Jack Temple Kirby all warmly endorsed both the suggestion of a reprint and my proposed introduction. In writing the introductory essay I found the comments of Professors Ayers and Honey especially helpful, although I recognize they may not agree with everything I say. Vernon Burton also graciously read and commented on a draft of the essay. The introduction presents my own interpretation of the STFU, but it is primarily a synthesis which rests on the work of others; the considerable debt I owe to the scholars who have produced fine research on Kester, the STFU, and their milieu should be obvious.

The text of Kester's book appears exactly as it first did in 1936. The photographs, however, are my own selection from the Southern Tenant Farmers' Union Papers and the Howard A. Kester Papers at the Southern Historical Collection at the University of North Carolina's Wilson Library. John White of the Southern Historical Collection provided me with indispensable aid in locating these photographs. Florida International University provided a travel grant for the photographic research. I thank Nancy Kester Neale, Howard Kester's daughter, for graciously granting permission for republication of her late father's work.

Finally, as always, Sybil Lipschultz has provided me with unstinting love, encouragement, and support throughout my work on this project.

Hannah, this one's for you.

Wilmette, Illinois
July 17, 1996

INTRODUCTION

The Southern Tenant Farmers' Union: A Movement for Social Emancipation

We face the future with all those who earn their bread by the sweat of their brow, who hate tyranny and oppression and who love justice, truth and freedom.

—Howard Kester and Evelyn Smith, "Ceremony of the Land," 1937

I

In August of 1937, Howard "Buck" Kester wrote his close friend and political confidant Socialist Party (SP) leader Norman Thomas to voice his concern about affiliation of the nascent Southern Tenant Farmers' Union (STFU) with the powerful new Congress of Industrial Organizations (CIO). Formed by black and white sharecroppers and tenants in Arkansas three summers before, the STFU now had the opportunity to obtain full legitimacy in the increasingly influential ranks of organized labor. Yet Kester warned that such a move might dilute the unique character of the sharecroppers' organization. "I think something can be said on behalf of the Union being something more than a union," he reminded Thomas. "It is besides this a movement. . . . The problem of tenants and the land, particularly here in the South, is one which lends itself to the development of a genuine emancipation movement which encompasses the whole of life," he concluded.[1]

This "movement for social emancipation," as Kester called it elsewhere, was defined by paradox.[2] In a segregated society, the STFU brought together white and black agricultural laborers in common cause. Part of a nationwide upsurge of industrial unionism, it championed agrarian values. Kester, a deeply religious man, was also a committed revolutionary socialist, as were most of the leaders of the STFU. The union emerged at a moment when, for the first time in a generation, social reformers could look to the national government for help; yet one of the pillars of New Deal reform, the Agricultural Adjustment Administration (AAA) proved to be the sharecroppers' nemesis. The southern tenants, sharecroppers, and day laborers who constituted the STFU's rank and file represented the very bottom of American society, yet they managed to attract the attention of Secretary of Agriculture Henry Wallace, First Lady Eleanor Roosevelt, and many other powerful members of the "eastern establishment." Finally, facing relentless retaliatory violence and terror, STFU leaders preached the principles of non-violence.

These contradictory tendencies no doubt undermined the staying power of the STFU, which had already seen its heyday pass by the close of the 1930s. Yet the capacity of America's "disinherited" to express hopes and dreams for a better world came alive again during the 1960s. The Student Nonviolent Coordinating Committee (SNCC), the United Farm Workers (UFW), Martin Luther King Jr.'s "Poor People's Campaign," and other multiracial movements for social justice expressed many of the goals first articulated by the STFU three decades earlier, and drew on many of the same sources for sustenance.

A handful of good books, including the memoirs of Harry Leland Mitchell, the union's long-time executive secretary and guiding force, have been written about the STFU, and the union merits several paragraphs in virtually every history of the twentieth-century South, American radicalism, agricultural history, or the New Deal (see the suggested readings at the end of this essay). Moreover, both Kester and Mitchell were notorious packrats; the STFU Papers and the Kester Papers, which together document in astoundingly rich detail the day-to-day history of the union, its organizers, its allies, and its rank-and-file members, are housed at the Southern Historical Collection in the Wilson Library at the University of North Carolina. The entire STFU collection and selections from Kester's papers are also available at many libraries on sixty reels of microfilm. Thus the STFU's historical record is hardly obscure. But for today's student, social activist, or interested citizen without the time to sift through the thousands of documents in these collections, there is no substitute for a contemporaneous narrative account of how this remarkable organization came to be, the social vision it championed, and the obstacles it faced.

In order to fill this gap, Howard Kester's "pamphlet" about the STFU, *Revolt among the Sharecroppers*, written during the second half of 1935 and published in March 1936, is here restored to print. Intended to stir sympathy (and raise badly needed funds) for the fledgling union, *Revolt* is one of the more remarkable documents of twentieth-century southern agrarian radicalism. It should be of interest to students of the cultural and political milieu of the 1930s, the New Deal, the rural South and its peculiar race relations, the labor movement, and the American left—or anyone interested in a still-neglected strand of radical American social thought. Unlike even the best histories or most reliable memoirs, *Revolt among the Sharecroppers* captures first-hand the immediacy and excitement of the establishment of the STFU and its first year of life at a time of enormous social upheaval and political ferment.

During the Depression decade, in no small part due to the efforts of the STFU to publicize their plight, southern sharecroppers and tenants of both races pricked the conscience of a nation long indifferent to the normally invisible problems of the rural poor. Organizations devoted to liberalism, to socialism, to socially responsible Christianity, or to civil rights—Kester's own Committee on Economic and Racial Justice, the Federal Council of Churches of Christ, Gardner Jackson's National Committee on Rural Social Planning, the Workers' Defense League, the League for Industrial Democracy, the National Association for the Advancement of Colored People (NAACP), the American Civil Liberties Union, among others—supported the sharecroppers' struggle for justice with protest and publicity campaigns, lobbying efforts in Washington, fund-raising, and contributions. If the 1930s began with Erskine

Caldwell's portrayal of the pathetic victimhood of impoverished rural southerners in *Tobacco Road* (1932), by the end of the decade President Roosevelt's Committee on Farm Tenancy and the federal agency it spawned, the Farm Security Administration (FSA), began to take the problems of farm tenancy and rural poverty seriously at the highest levels of the New Deal.

In the intervening years, the widespread interest in the sharecroppers' story generated an entire bookshelf of written material. Norman Thomas's own political tract, *The Plight of the Share-Cropper,* appended to the local Socialist Party's survey of plantation conditions in the Arkansas Delta, first brought the rural social crisis in that region to the nation's attention. In Thomas's wake came rural sociologists who produced academic studies and foundation-funded reports bemoaning the "Collapse of Cotton Tenancy," as the most influential of these studies was called. Writers of "proletarian" fiction incorporated into their novels the interracial efforts of the sharecroppers to organize, and Depression-era travel writers made the obligatory stop in the Delta. Caldwell himself, in collaboration with FSA photographer Margaret Bourke-White, published *You Have Seen Their Faces* in 1937; and James Agee and Walker Evans offered their own belated antidote to *Tobacco Road* in their effort to restore dignity to white sharecropper families in *Let Us Now Praise Famous Men* (1939). Yet among this outpouring of work, Kester's book stands out as the only sustained effort to recount the STFU's story by someone engaged in the day-to-day struggle to build the union from the bottom up.

Howard Kester

Like many socially active young people during the Depression years, by the time of his involvement with the STFU at the age of thirty, Kester had become a committed socialist. Yet despite his fervent belief in class struggle and the desirability of doing away with capitalism as the nation's economic and social system, "there was more of Jesus and Jefferson than Marx and Engels in his vision," as his biographer perceptively remarks. The "Jesus" came from his highly religious upbringing; indeed, from an early age Kester assumed that he would enter the Presbyterian ministry and pursued a religious education and a commitment to Christian service.

His religious radicalism derived from his growing sense that the organized church restricted the biblical ethic of Christian brotherhood by both race and class. Because of his social activism, the Presbyterians never ordained him, yet this only strengthened his attachment to an ethical religious imperative that often struck other Socialists, including H. L. Mitchell, as somewhat out of step with revolutionary politics. Kester himself remarked that "when I joined the Socialist Party I was attacked for my religion and when I wished to become a minister my religion was attacked because of my socialism."[3]

The "Jefferson" in Kester similarly sprang from a disjuncture between the ideals and social reality of his youth. Born in 1904, Kester spent his early years in a rural community on the outskirts of Martinsville, Virginia, a small town where his father worked as a tailor. Few of the disruptive forces of modernization then sweeping across the South impinged upon the young Kester's romantic sense of an agrarian social life defined by personal and direct economic relationships. But at the age of twelve his family relocated to Beckley, a coal-mining town in West Virginia. Here rapid economic growth and industrial development spurred by powerful corporations cracked open a growing gulf between the working and the middle class. Here too modernization exacerbated racial conflict, as thousands of African Americans were drawn to the West Virginia coalfields only to meet white hostility. Life in West Virginia, home to violent miners' strikes and a resurgent Ku Klux Klan (which his father joined), provided Kester with a first-hand education in class and race inequality.

This was an education he put to good use during the 1920s as a Christian activist for the Young Men's Christian Association (YMCA) and the Fellowship of Reconciliation (FOR). While both of these organizations emphasized pacifism, Kester's activism in the cause of peace led him and a handful of other young southern Protestants into interracial fraternity for the first time. The ethical commitments to both non-violence and interracialism so central to Kester's work with the STFU find their source in his role as a southern youth organizer for these Christian pacifist groups.

In the atmosphere of the South in the 1920s the decision merely to organize and attend interracial gatherings under the auspices of the YMCA or FOR marked a radical break from the region's social orthodoxy on race. For nearly three decades, legally enforced segregation in all aspects of southern life had held sway across the region, having received the U.S. Supreme Court's stamp of constitutional approval in 1896 with the *Plessy v. Ferguson* ruling. The political gains made by African Americans in the 1860s and 1870s during Reconstruction had long since been rolled back, and a maze of legal and financial barriers had been thrown up to prevent blacks from registering to vote. By the turn of the century, the southern electoral rolls had been purged of the vast majority of their African American voters. For blacks who showed too much desire for economic independence, demanded political rights, or dared to cross the color line, summary execution by a mob—lynching—was often the punishment. Over three thousand lynchings of African Americans were recorded between 1889 and 1932, and many others went unreported.

It was highly unusual for a white person to dissent from this state of affairs other than to show anything more than the most paternalistic sympathy for the other race. For Kester, interracial encounters in an egalitarian atmosphere proved both eye-opening and liberating. "Negroes seemed to trust me and believe in me," he remarked years later, and "the problems of the Negro became my problems." Because of his casual flaunting of the color line, Kester was often mistaken for a black man "and treated as such." He and his wife Alice paid a price for their rejection of the caste system, however, as they were ostracized by their families who unquestioningly accepted the southern racial code.[4]

The Radical Gospel

It was with this personal ethic of interracial solidarity firmly in place that Kester encountered the radicalizing influences of the Great Depression and advanced education at, of all places, Vanderbilt University in Nashville, Tennessee, where he enrolled in divinity school in the late 1920s. Consideration of the southern contribution to social and political thought in the Depression era usually conjures up the "Nashville Agrarians." This handful of Vanderbilt University literature and philosophy professors challenged the regnant values of secularism, industrial progress, and modernity with a conservative vision of traditional values rooted in the southern community life of agrarian society. Kester and several other young men destined to take up the sharecroppers' cause and make a contribution to southern radicalism attended Vanderbilt during the reign of the Agrarians but remained untouched by their romantic yearning for a lost southern past. Instead, under the influence of Alva Taylor at Vanderbilt's divinity school, Kester, Ward Rodgers, and Claude Williams developed a far more radical critique of the ills of the modern South and embraced an ideal of dramatic social transformation. Like their conservative counterparts, Taylor's students (and Kester in particular) sought a renewal of spiritual values, questioned the material benefits of capitalist and industrial society, and believed in the viability of rural life; but they did so from a socialist perspective.

For Kester, who struggled to reconcile his increasing commitment to social justice with his profound religious beliefs, Alva Taylor provided an indispensable link to a renewed Social Gospel. Taylor had long been associated with this school of Christian thought, which since the 1890s had sought to bring the ethical teachings of Jesus to bear upon the social problems of modern industrial society. "When science gives the technique and the Church gives the social passion," proclaimed Taylor, "we will possess the power to make the world over into the Kingdom of God."[5] Influenced by Taylor, Kester soon embraced a "prophetic" tradition of religious thought which measured social justice against an ethical standard derived from the Old Testament prophets. By the time he received his divinity degree in 1931, Kester thought that Jesus "definitely recognized the class struggle and set his face against the oppressors of the poor, the weak and the disinherited."[6]

Under Taylor's tutelage, Kester, Rodgers, and Williams helped forge what historian Anthony Dunbar calls a "radical gospel" tailored to the conditions of their native South. Along with Christian socialists James Dombrowski and Myles Horton, who founded the Highlander Folk School and nurtured a generation of southern labor union and civil rights advocates, the Vanderbilt radicals believed that spiritual renewal went hand in hand with social transformation. Like Kester, the prophets of the southern radical gospel grew up in the small-town or rural South and witnessed the impact of modernization on a bucolic world; like Kester they found themselves drawn to religious training in mainline Protestant churches; and like him, with the coming of the Great Depression they dedicated their lives to helping the dispossessed croppers, textile and mine workers, and unemployed of the South, who bore the brunt of the economic collapse. Moreover, unlike the Agrarians, Taylor and his disciples firmly believed segregation was incompatible with Christianity, and tended to practice what they preached in racial matters.

Kester's own engaged radical gospel found its first expression in his moral witness and material support for striking miners in Wilder, Tennessee in 1932–33. Only one hundred miles east of the civilized enclave of Vanderbilt, Kester encountered a feudalistic company town in which the Fentress Coal and Coke Company maintained an iron grip on all aspects of its employees' lives. When the already impoverished miners faced a unilateral 20 percent wage cut in June 1932, they went on strike. The ensuing year-long violent conflict claimed the life of the mine union's leader, Barney Graham, who was shot in the back by a hired company thug. The dignity and courage shown by the striking miners in the face of overwhelming odds, and the preponderance of state and corporate power which led to their defeat, made it clear to Kester which side he was on. Profoundly influenced by his encounters with Socialist Party leader Norman Thomas and the radical theologian Rheinhold Niebuhr (who contributed a preface to *Revolt among the Sharecroppers*), Kester had joined the party in 1931 and helped organize its Tennessee branch. Nevertheless, the showdown between the powerful and the powerless in the coalfields further radicalized him, pushing him to the far left wing within the wide spectrum found within the Socialist movement of the 1930s.

Thus armed with his militant interracialism, his fervent belief in the necessity of class struggle, and his philosophy of Christian radicalism, Kester threw himself into the myriad struggles of his era. He worked in an official capacity as the field secretary of a new organization created by Niebuhr, the Committee on Economic and Racial Justice. In practical terms this gave him the freedom to travel around the South in an effort to promote the Socialist Party, encourage contact across racial lines, help organize agricultural and industrial workers, and investigate specific instances of racial injustice.

This latter task brought him into direct confrontation with one of the unique horrors of the South, the lynching of black men accused of sexual assault on white women. The 1930s saw both an increase in the number of lynchings across the South and new political openings for the NAACP's long-standing campaign against these atrocities. When in the fall of 1934 a mob in northern Florida brutally tortured, lynched, and mutilated a young black man named Claude Neal accused of raping and murdering a white woman, the NAACP sought to publicize widely the case in an effort to secure federal anti-lynching legislation. Walter White, secretary of the NAACP, asked Kester to investigate the Neal lynching and prepare a report. This was a dangerous assignment, which no black person could undertake. Posing as a Methodist book salesman, Kester traveled to north Florida where his southern accent and white skin provided him with cover, and many whites happily regaled him with accounts of the lynching in gruesome detail. His report, *The Lynching of Claude Neal,* was distributed by the NAACP and helped shift public sentiment, even in the South, against the practice of vigilantism.

Despite this publicity, the anti-lynching bill before Congress failed, filibustered to death by the southern Democrats who formed a key component of Roosevelt's New Deal coalition. For Kester, Neal's lynching reflected the economic and social tensions generated among the poor of both races by the depression. But the failure of the evidence he gathered to secure federal protection for African Americans dramatically illustrated the limits of the NAACP's emphasis on legislation, litigation, and publicity. "Increasingly we shall have to direct our attention and efforts to the exploited and disinherited whites and negroes among whom the [racial] conflict now rages," he confided to a friend at the end of 1934.[7] A recently created union of black and white sharecroppers in the Arkansas Delta offered just such an opportunity.

Grass-roots Socialism

Shortly before he had undertaken his assignment for the NAACP, Kester had heard from fellow socialist H. L. Mitchell, whom he had met at an SP conclave in New York the year before. Kester had come to his socialism along an intellectual and even spiritual path; the secular Mitchell, by contrast, was a product of what historian James Green has called "grass-roots socialism," which had an especially strong historical presence in early twentieth-century Arkansas, Louisiana, Oklahoma, and Texas. Like Kester, Mitchell had grown up in the small-town South during the first two decades of the twentieth century, but in west Tennessee, an area far more rooted in the plantation economy than Kester's native western Virginia. While Kester pursued his religious education, the young Mitchell worked as a newspaper deliverer, bootlegger, and sharecropper. Galvanized by Socialist Eugene V. Debs's 1920 presidential campaign, Mitchell educated himself with the *Little Blue Books* published by the Girard, Kansas, socialist newspaper, *Appeal to Reason*. In 1928 Mitchell relocated to Tyronza, in the Arkansas Delta, where he set up a dry-cleaning business and talked socialism to anyone who would listen.

By 1932 he and one of his converts, gas station owner Clay East (their adjacent businesses in Tyronza were known locally as "Red Square") had chartered a local branch of the Socialist Party and invited Norman Thomas to the Delta to address the comrades they had collected and conduct research for the SP's pamphlet on cotton tenancy, *The Plight of the Share-Cropper*. As unlikely as it may seem today, socialism was not entirely out of place in rural Arkansas. Mitchell and East easily tapped the dormant sentiments of southwestern agrarian radicalism, dating back to the 1890s. In the waning years of the nineteenth century the Populists had challenged the power of banks, railroads, and speculators in an effort to preserve the land of this region for the use of small farmers. Between 1900 and 1920 the socialism of Eugene V. Debs had fired the imagination and borne the aspirations of workers and farmers in Arkansas, Louisiana, Oklahoma, and Texas; indeed, these four states together gave Debs over eighty thousand votes in the 1912 presidential election. The Socialist Party in Oklahoma alone boasted 961 locals, twelve thousand members, and over one hundred elected officials in 1914. Radical unions, including the Industrial Workers of the World–led Brotherhood of Timber Workers in Louisiana, and the United Mine Workers in western Arkansas, had a presence in the region. And the dozens of small socialist newspapers (like the *Appeal to Reason* discovered by a young Mitchell) and publishing houses in these and neighboring states kept the Socialist gospel alive even during its political demise in the 1920s.[8]

Together these organizations and institutions had resisted the spread of the modern capitalist economy into a region that had once promised small farmers and migratory workers personal independence and economic opportunity. As absentee landlords, timber companies, banks, and coal-mining concerns reshaped the region's class relations, these men and women found themselves sinking into the status of dependent proletarians. In the face of this unwelcome transformation, they created a dense radical tradition which socialists easily revived in the 1930s when working people cast about anew for a political response to the depression.

H. L. Mitchell, as both an inheritor of this tradition and one of its proselytizers, was ideally placed to carry it on. Together he and Kester represented two very distinct strains of American socialism, a renewed Social Gospel and

One of the most prominent supporters of the STFU was American Socialist Party leader Norman Thomas, pictured here in 1937 addressing an STFU rally in tiny Birdsong, Arkansas. A former minister and a spellbinding orator, Thomas urged H. L. Mitchell and Clay East to organize a sharecroppers' union when he first visited the Arkansas Delta in early 1934. Kester's attack on the economic assumptions underlying the New Deal's agricultural recovery programs owed a large debt to Thomas's insistence that the American economy should be reorganized to meet human needs, a view Thomas put forth in his presidential campaigns of 1932 and 1936. H. L. Mitchell also credited Thomas with inspiring his own interracialist views. Southern Historical Collection, Library of the University of North Carolina at Chapel Hill.

the radicalism of the southwestern frontier. Their collaborative effort to build a union of agricultural workers exemplified the powerful synergy generated by this unusual hybrid. Mitchell wrote Kester in August 1934, several weeks after the formation of the STFU, to invite him to "attend one of our meetings with black and white sharecroppers and hear their fundamentalist preachers preach revolution to them."[9] The shrewd Mitchell obviously had a keen sense of the Christian radical's affinity for the interracial movement he hoped to plant in the inhospitable soil of the Arkansas Delta.

Debt in the Delta

When Howard Kester crossed the Mississippi River from Memphis into northeastern Arkansas for the first time in late 1934 he entered a land of stark economic inequality. The roughly ten thousand square miles of fertile bottomlands squeezed between the White and the Mississippi Rivers west and north of Memphis were characterized by highly concentrated land ownership, a single-minded devotion to cotton cultivation, an African American majority, and an extremely high rate of farm tenancy. In the STFU strongholds of Poinsett, Crittenden, and St. Francis Counties, for example, 80, 88, and 95 percent respectively of all farmers were tenants rather than owners in 1930.[10]

The plantation society of the Arkansas Delta was no relic of the Old South, however. Instead, it was a product of the post–Civil War spread of large-scale cotton production west of the Mississippi. The once-forested land in the Delta had only been brought into production in the last decade before the Civil War and remained only sparsely settled in the antebellum years. After the war, as timber companies cleared the land and levees tamed the flood-prone rivers, the modern economic forces of capitalism reshaped the region's social relations. The influx of the dispossessed of both races and the increasing precariousness of the small farmer in the face of high taxes and low cotton prices in the closing years of the nineteenth century led to a dramatic transformation in the land tenure system of the Delta. In 1880, three-fourths of the farmers in Poinsett county, birthplace of the STFU, owned their own land while only one-fourth of them had to work for someone else. By 1920 this proportion had been reversed.[11]

With some justification, Kester used the term "sharecropper" to cover a wide variety of tenure arrangements between landowners and laborers in the plantation South. A handful of tenants paid cash or a fixed amount of cotton as rent to the landowner. This represented a relatively exalted—and thus rare—position on the agricultural "ladder" (a notorious misnomer, since it implied the possibility of gradual progress upwards). Although still supervised by the landlord, these families owned their own farm tools and mules, supplied the seed and fertilizer, managed the land they worked, maintained some independence, and reaped the benefit of the crop once they paid their rent. Below these "cash tenants" came the "share tenants" and then the "share croppers," who together made up the majority of southern farmworkers on the eve of the depression. These tenant families brought far less of their own equipment to the "partnership" with the landowner and paid from one-fourth to one-half of the

In common with many of his contemporaries, Kester discovered that images often spoke louder than words. Consequently, he sought to document the conditions he observed in the Arkansas Delta with his camera as well as his pen. This photograph of typical sharecropper dwellings is just one of many pictures Kester took of the primitive living conditions he found when he traveled to northeastern Arkansas in 1935. Kester Papers, Southern Historical Collection, Library of the University of North Carolina at Chapel Hill.

crop—regardless of its value—in rent. Rather than directing their own labor, the croppers near the bottom of the ladder were beholden to the landowners' "riding bosses" in the fields. The further down the agricultural ladder one went, the more dependent one became upon the landlord for management, housing, tools, seed, fertilizer, mules, and everything else, and the more of the crop went to the landowner for both rent and to pay back the advances made at the beginning of the season. Below these tenants and croppers came the day laborers, often paid in scrip redeemable only at the plantation commissary. This rural proletariat had absolutely no claim to the crop they helped harvest. Kester and many other observers quite eloquently testified to the impact of this plantation system, and the abject dependence it required, on the health, diet, education, and material and moral condition of the impoverished rural people whose labor made the cotton planters rich.

As Kester notes, at bottom it was the debt and credit system that gave the landholders their powerful leverage over the daily lives of their sharecroppers. On March 1 of each year the sharecroppers would gather at plantation commissaries all across Delta (and the South) to receive their "furnish" of supplies for the coming planting cycle. This initial debt accumulated interest over the next ten months, sometimes at a rate as high as 40 percent. From March until mid-summer the men, women, and children in sharecropper families would plant, weed, and chop the cotton; from September until the end of the year they would stoop over to pick it. Any additional supplies could be had from the plantation commissary at inflated prices, with interest, or paid for in the undervalued scrip (in Arkansas known as the infamous "doodlum books" mentioned by Kester) issued by the planters and good only at the commissary. At settlement time, around Christmas, the planter ginned the cotton, took his share, deducted the year's charges from the cropper's share of the proceeds of the crop, including principal and interest on the originally "furnish," and paid the cropper whatever cash remained. Needless to say, with a little creative calculation on the landlord's part this often amounted to very little or nothing at all. The *STFU News* recounted this piece of sharecropper folk humor:

> **Teacher:** "If the landlord lends you twenty dollars, and you pay him back five dollars a month, how much will you owe him after three months"?

Student: "Twenty dollars"
Teacher: "You don't understand arithmetic"
Student: "You don't understand our landlord!"[12]

On a more serious note, the scholars who researched *The Collapse of Cotton Tenancy* estimated that by the early 1930s more than one-third of southern tenants carried debts of more than a year's standing.

The long-term consequences of this so-called "crop-lien" system on the South are well known: widespread poverty with no hope of advancement among landless workers of both races, a debilitating dependence on cotton as the sole crop, a fatal lack of incentive to mechanize or diversify southern agriculture, an extreme maldistribution of resources, and an equally skewed distribution of political power to insure the system's stability. The static appearance of this social order was deceptive, however. Sharecroppers disproportionately paid the social and economic costs of floods, drought, and the boll weevil pest—the trinity of natural disasters that plagued the Delta in the decades prior to the Great Depression. When the price of cotton declined precipitously, as it did in the human-made disaster of the early 1930s, it threw the already badly frayed plantation economy into a terminal crisis. For tenants and sharecroppers, perpetually in debt to their landlords, plummeting prices meant that the proceeds of their share of the cotton crop progressively covered less and less of the debt they incurred to grow it. If a five-hundred-pound bale of cotton brought in three hundred dollars less in 1932 than it had in 1931, the landlord recouped his loss through the repayment of last year's "furnish," at last year's prices plus interest. Put another way, the cropper inadvertently bore the risk of the commodity market in cotton, and when the bottom dropped out he was the loser. Thus the debt and credit system of plantation agriculture made the sharecropper especially vulnerable to the Great Depression, but when it came the crisis made his deeper and long-standing grievances that much more sharply felt.

Impact of the AAA

Central to the sharecroppers' grievances as well was the impact of the New Deal agricultural recovery programs on the lives of those who worked the land. In March 1933, at the height of the Great Depression,

President Franklin D. Roosevelt took office with a mandate to stem the tide of economic crisis. In the three-and-a-half years since the stock market crash, all across America banks had closed their doors, wiping out personal savings; industrial production had slumped by more than 50 percent; unemployment had swelled to twelve million, 25 percent of the workforce; and the price of cotton had sunk to six cents a pound, one-third of its 1929 price. Thirteen million bales of cotton remained unsold, an enormous surplus that threatened to depress the market further.[13]

Consistent with their overall interpretation of the causes of the depression, Roosevelt's "Brain Trust" of economic experts reckoned that overproduction had undermined agricultural prices. The solution, embodied in the Agricultural Adjustment Act, was to restrict agricultural output. Part of Roosevelt's "First Hundred Days," the sweeping legislative program designed to promote economic recovery, the act was passed on May 12, 1933. The South's cotton crop was already in the ground. So the AAA, charged with administering the new agricultural recovery program, paid farmers—that is, farm *owners*—to plow up a portion of their crop, a process marvelously evoked in *Revolt among the Sharecroppers*. In Arkansas that summer, mules plowed under one-quarter of the cotton crop. The following season, the AAA began paying farm owners to reduce the acreage that they planted in cotton in order to drive up the price.

Perhaps the most significant contribution of *Revolt among the Sharecroppers* to the discourse of 1930s reform is Kester's powerful critique of the AAA program. The socialist was dismayed both by the AAA's practical effects on those who worked the land and its basis in an economics of scarcity. Today we tend to associate the New Deal's pioneering federal intervention in the nation's economy with "liberalism" (positively) or "big government" (negatively). But it is instructive to recall that dissenters on the left voiced concern at the time that programs like the AAA stimulated economic recovery for the benefit of the powerful and wealthy while exacerbating the desperate plight of the powerless and poor. Most historians of the New Deal now recognize this negative impact of Roosevelt's policies, particularly in the so-called First New Deal of 1933–35. Kester's pamphlet reminds us, however, that just such a critical perspective had vigorous advocates from the inception of the agricultural recovery program.

The simple procedure adopted by the Department of Agriculture of making out the AAA checks payable to the landowner had enormous social consequences, as Kester's account suggests. Quite predictably, given the imbalance of social power in the plantation South, many planters simply kept the entire AAA payment for themselves, refusing to pass on the tenant's share (meager in any case). In order to keep up appearances, some planters claimed their tenant's payment as a fulfillment of their debt. Moreover, the planters' removal of cotton acreage from production naturally made a certain number of tenants and sharecroppers superfluous. Under the terms of Section 7 of the AAA's Cotton Contract (reproduced in full by Kester), tenants were permitted to remain on the land, but as Kester demonstrates this was really at the landlord's discretion. Flush with federal dollars, the planters evicted their sharecroppers and replaced them with day laborers whom they could pay in cash, leaving tenant families literally out in the cold. Indeed, as *The Collapse of Cotton Tenancy* showed, in many cotton counties federal relief expenditures rose in tandem with AAA benefit payments. Moreover, the infamous clause of Section 7 that allowed tenant dismissal for being "a menace to the welfare of the producer" was unhesitatingly applied by landowners to members of the STFU, who, in the eyes of Delta cotton planters, certainly constituted the greatest menace since the Union Army.

Writing in the NAACP's magazine, *The Crisis,* in June 1935, socialist preacher and STFU activist Ward Rodgers pointed out that "the most important grievances of the STFU grow out of [Secretary of Agriculture Henry] Wallace's program."[14] Sharecroppers and their advocates, both inside and outside the government, challenged the uneven division of AAA payments, the outright denial of the sharecropper's portion, and the epidemic of evictions, but to little avail. A handful of attorneys inside the AAA sought to shore up the rights of tenants to their share of acreage reduction payments but came up against the traditional mores of the plantation South and its powerful Democratic representatives in Congress. Roosevelt felt it necessary to placate these southern congressmen and senators in order to preserve the political coalition that made the New Deal possible. When AAA lawyer Alger Hiss suggested making out benefit checks jointly to landowners and their tenants, an outraged South Carolina Democratic senator "Cotton Ed" Smith charged into his office exclaiming in disbelief "you're going to send money to my niggers instead of to me?"[15]

The AAA Cotton Section's compliance division was staffed by people like Senator Smith: southern cotton planters, their friends, family, and political cronies, especially at the local level where sharecroppers were denied representation on AAA county boards. In 1934, for example, by September 1,457 complaints had been filed with the AAA about the administration of the cotton program, 477 from Arkansas alone. Of these only 21 (and only 11 in Arkansas) resulted in canceled contracts. The historian of southern agriculture Pete Daniel aptly remarks that "the [AAA] complaint machinery fed the victim to his oppressor."[16]

By February 1935, Hiss, Jerome Frank, Gardner Jackson, and the other defenders of sharecroppers' rights at the AAA found themselves "purged" from the agency, victims of the enormous influence wielded by southern Democrats like Arkansas Senator Joseph Robinson, whose support in Congress appeared essential if the New Deal were to move ahead. At the same time, an explosive investigative report on conditions in the Arkansas Delta reached the desk of chief AAA administrator Chester Davis. The "Myers Report," as it came to be known, was the result of an investigation by Mary Conner Myers, a genuinely impartial observer. She was shocked by what she found when she traveled to Arkansas at the behest of Jerome Frank, head of the AAA's legal section. The affidavits she collected, USDA correspondence, and newspaper reports suggest that Myers confirmed the STFU's reports of desperate poverty, pilfered AAA payments, evictions, and relentless repression. We may never know what she said in her report, however. Chester Davis immediately suppressed it, and to this day no researcher has been able to locate a copy of this document in the papers of an otherwise exceptionally well-documented government agency. With the suppression of the Myers report, STFU leaders claimed at the time, "the cropper was left to his fate."[17]

The STFU and Interracialism

No aspect of the STFU's remarkable history has drawn more attention than its mostly successful effort to bring together whites and blacks in common struggle. The NAACP, normally engaged in challenging racial discrimination by organized labor, embraced this "unprecedented mixed union of white and colored sharecroppers" with enthusiasm. Through his association with Walter White, Kester himself served as an indispensable link between the respected civil rights organization and the Socialist-led STFU. Kester touted the union's interracial record when he addressed the NAACP annual convention in June 1935, shortly before he began to write *Revolt among the Sharecroppers*. "I want the NAACP to become sharecropper conscious," he declared. "I want the NAACP to become a sounding board upon which the cries of these exploited workers can be sounded." For his part, Walter White helped subsidize publication of Kester's pamphlet, and the NAACP provided funds to the cash-starved STFU when it could. This sort of collaboration marked one of the high points of antiracism in the 1930s, when organizations of the left and more mainstream civil rights advocates together recognized the necessity of bringing together the working classes of both races. "Together they must be saved or together they must be lost," as Kester told the NAACP delegates that day.[18]

In a sense the union's racial egalitarianism was far more radical than its initial economic program, which, after all, at first merely sought fair enforcement of the AAA contracts and the extension of the rights granted to workers by the New Deal to agricultural laborers. But open violation of the South's laws and customs of segregation threatened a very real revolution in race relations. As Kester and other southern radicals were well aware, broaching the color line and undermining southern racism meant far more than a simple rearrangement of the social relations between whites and blacks; this they scornfully left to what they saw as the paternalistic liberals of the Commission on Interracial Cooperation (the "Atlanta School" so derisively dismissed in *Revolt among the Sharecroppers*). Instead, in bringing the "disinherited" of both races together, the STFU sought to overturn the entire southern economic and political structure of which racism was an integral part.

The inequities of the plantation system, which by the 1930s affected more whites than blacks in the South, were inseparable from the efforts to preserve white supremacy that characterized southern politics after the overthrow of Radical Reconstruction in the 1870s. During the ten years following the Civil War, emancipated slaves had forged an alliance with some whites to create the southern wing of the Republican party. At their best, the "Radical Republicans" simultaneously defended the economic rights of small landholders and agricultural workers and the civil and political rights of African Americans. The decades following the destruction of the southern Republican party and its program in the 1870s saw rising rates of landlessness, tenancy, and debt for blacks and whites alike. By the waning years of the nineteenth century this went hand in glove with an increase in lynchings, the spread of convict leasing, the hardening of segregation, and the disfranchisement of African American voters.

The consequences of these late nineteenth-century developments for African Americans are clear: relegation to the bottom of a rigid caste system which they challenged at their peril, a complete lack of political power, and a deep and crushing poverty from which there was little chance of escape. Black men driving the dusty roads of the rural South in the twentieth century were expected to avoid passing a white driver no matter how slowly they drove lest they soil the white's vehicle; to express a desire for economic independence, let alone to register to vote, endangered one's life and property; black tenants who fell in debt to their landlord violated the law if they attempted to quit the plantation or work for someone else. Those who broke their contracts and were caught faced an unenviable choice of working on the chain gang or paying a fine, the cost of which would be borne by the planter and added to the tenant's spiraling debt.

Poor whites, however, appeared to be the partial beneficiaries of black subordination, and greeted with great enthusiasm the campaigns of the race-baiting politicians who promised to enforce the caste system. But in the long run they paid a steep price, for the South's economic dependence on cheap, degraded labor and the political system designed to preserve it made them its victims too. By the 1930s, if not long before, the habit of racial control and exploitation of the plantation "carried over to the white tenant and cropper" even while "the poor white connives in this abuse of the Negro," wrote the authors of *The Collapse of Cotton Tenancy*. Visitors to the Arkansas Delta frequently observed that, regardless of race, "the word sharecropper itself has come to be used as a term of contempt," almost interchangeable with "nigger." The South, as Kester noted, "withheld economic, civil, and political privileges from white men" in an effort "to deny them to colored men."[19]

With most blacks and eventually many poor whites kept from the voting rolls by the poll tax, literacy tests, or outright coercion, a handful of powerful elites, representing the interests of planters, timber barons, and coal-mine owners, came to control the southern wing of the Democratic party—and in the South, by the twentieth century, this was the only party.

"The most Democratic part of the nation," remarked rural sociologist Arthur Raper, "is perhaps the least democratic part of the nation." In Congress, this wing of the Democratic Party controlled the purse strings of Roosevelt's New Deal coalition and often exerted their influence to check any reform that might threaten genuine social change in the South.[20]

Despite Kester's claim to the contrary, the STFU was hardly the first organization in the post-Reconstruction South to heed Populist Tom Watson's famous dictum to the poor of both races that "you are kept apart that you may be separately fleeced of your earnings."[21] Indeed, some historians have argued that the consolidation of white supremacy between 1890 and 1910 came in direct response to the threat posed to the southern Democrats by Watson's People's Party and its appeal to the economically pressed of both races. To this should be added some remarkable attempts on the part of trade unions to cross racial lines during the same era. In Birmingham, Alabama, the industrial heartland of the Deep South, between 1890 and 1908 black and white coal miners joined together in the United Mine Workers in an effort to thwart the power of the city's coal barons. These miners worked together, drank together, and went on strike together. In New Orleans, where black and white stevedores loaded the South's cotton crop onto waiting ships, the Dock and Cotton Council led an interracial general strike in 1894 and continued to bring black and white workers together in the same labor organization into the twentieth century. And in the isolated piney woods of east Texas and western Louisiana the radical Industrial Workers of the World organized lumberjacks and sawmill operatives into the militant, interracial Brotherhood of Timber Workers. These were exceptions, to be sure, but they suggest that when economic conditions made racial cooperation a necessity, black and white workers might momentarily put aside whatever racial animosity they harbored and work together in their common interests.

This did not mean that whites abandoned their prejudices or blacks their mistrust when they joined an interracial union. Certainly for Kester racial fellowship was an end in itself, but as an organization the STFU, much like the People's Party before it, embraced interracialism as a *tactic* as much as a principle. In the Delta town of Marked Tree, for instance, the STFU initially established two locals, one for whites and one for blacks, "in conformance with the prejudice of the people."[22] But it was only a matter of time before members of each local began to attend the other's meetings in order to coordinate their activities. The Marked Tree local soon grew into an interracial organization of eight hundred members, led by a charismatic African American preacher, E. B. McKinney, who later became vice-president of the STFU. "That nigger's got more sense than any white man here," whites would confide to Mitchell when McKinney took the platform at these meetings.[23]

Because of the decentralized nature of the STFU and the small size of its locals—by 1937, for example, St. Francis County alone had over fifty locals which together consisted of fewer than forty-five hundred members[24]—isolated unmixed locals of blacks and whites did exist. But where blacks and whites worked and lived in close proximity they joined the same organization. And in the STFU as a whole, regardless of the composition of its locals, "there are no 'niggers' and no poor white trash," insisted Mitchell. "These two kinds of people are all lined up with the Planters. We have only Union men in our organization."[25]

It is obvious from Kester's account that blacks and whites recognized the practicality of joint action for different historical reasons, rooted in their distinctive experiences with the caste system. Hanging over the entire enterprise of building the STFU was the ghost of the "Elaine massacre," a racial pogrom touched off in 1919 when black sharecroppers in the southern part of the Arkansas Delta organized a "Farmers and Farm Laborer Association." Emboldened by the leverage provided by the mass exodus of African Americans to northern factories during World War I, this all-black organization hoped to boost the price sharecroppers and cotton pickers received for their cotton. Seen by whites as insurrectionary, this mild attempt at rural collective action was brutally suppressed, with dozens of blacks shot down and the ringleaders thrown in prison. One of the surviving farmers involved, Isaac Shaw, attended the founding meeting of the STFU in 1934 and reminded both races of the fatal consequences of an exclusively black union. As Ward Rodgers noted the following year in *The Crisis,* interracialism was the "best insurance against a race riot because the white sharecropper would be the first to be called by the planters to lynch leaders of a 'Negro Union'." For their part, while whites refused to entirely abandon their "sense of superiority," they grasped the futility of an all-white union: the planters would simply evict them and replace them with black tenants.[26]

Not only did blacks and whites have different reasons for joining an interracial union, but they also brought to the cause very different strengths. Long thrown back onto the resources of their community, African Americans had extensive experience with "mass organizations" in their churches, their fraternal orders, and their community self-help traditions. "My people are an emotional, and a *Group Conscious People,*" a black STFU organizer wrote to H. L. Mitchell, "and 98% of them can be reached in the churches." In contrast, white sharecroppers had little such

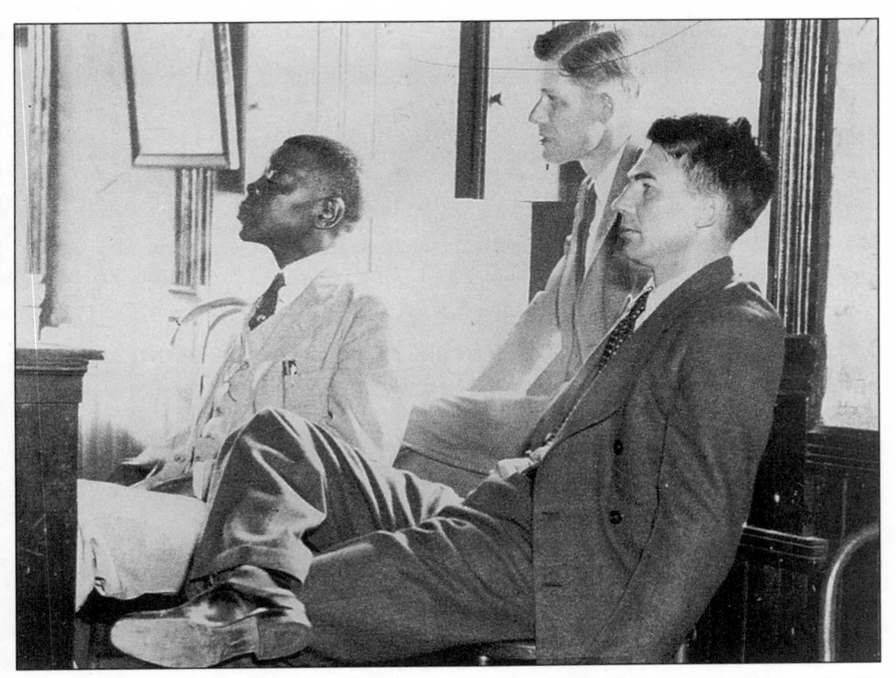

Presiding at a cotton choppers' strike meeting in May 1936, from left to right are E. B. McKinney, H. L. Mitchell, and Howard Kester. McKinney, vice-president of the union, was an articulate and highly race-conscious Baptist preacher. At one time a follower of Marcus Garvey's black nationalist Universal Negro Improvement Association, McKinney accepted the necessity of cooperation with whites but feared that African Americans reaped few of the benefits of interracialism. Mitchell, executive secretary of the STFU for most of its existence, brought to the union the tradition of agrarian socialism that had once swept the southwest. Kester, while consistently active in the STFU and an ex-officio member of its National Executive Committee, never held an official union position. STFU Papers, Southern Historical Collection, Library of the University of North Carolina at Chapel Hill.

collective experience and seemed far more individualistic, "ready to burn barns and lynch planters," in the words of one internal STFU report.[27] Although the organization struggled to contain such impulses, white sharecroppers offered the sort of credible threat of self-defense that was far too risky for blacks living in the Delta to brandish.

One of the most significant examples of this dynamic is strangely passed over in Kester's account. When a black preacher and STFU organizer, C. H. Smith, was arrested in Crittenden County the union attorney brought dozens of white sharecroppers with him to court. Faced with their hostile stares, the judge dismissed all charges against Smith. Kester tells this story in *Revolt among the Sharecroppers* but neglects to point out that this example of whites standing up for blacks proved a major demonstration of interracial solidarity for the STFU. Kester's own belief—or hope—was that "non-violence" would play "a dominant and all-important role in the struggles" of the STFU, as he wrote to the Fellowship of Reconciliation, but even he recognized that for most sharecroppers pacifism, like interracialism, remained a tactical rather than a philosophical position. Armed resistance to planter repression would merely provoke a bloodbath; "if we had fought back," Mitchell remarked years later, "it would have been our negro members who suffered most."[28]

The STFU and the "Union Gospel"

The significance of the STFU's commitment to interracialism lay in the hope that a tactical alliance across racial lines would breed a mutual respect and interdependence that would eventually lead to more genuine repudiation of the color line. The material basis for this alliance was a common economic plight rooted in the plantation economy; but if there was a cultural commonalty between the races for the STFU to tap it existed in the spiritual realm. Controversial bearers of the "radical gospel" like Ward Rodgers, Claude Williams, and Kester himself struggled to find a pulpit in southern Protestant denominations usually content to serve conservative middle-class communities. They finally discovered their true calling amongst the poor sharecroppers of the Delta. The established church in the region, personified by Methodist clergyman Abner Sage of Marked Tree, ministered to the needs of the well-to-do and aligned itself with the planters and their allies in social and economic matters, as Kester charged. The sharecroppers, black and white alike, spurned by the church, developed their own religious life, and found solace in the stirring oratory of backwoods preachers, E. B. McKinney among them. The combination of this lay preacher tradition with the organizational abilities and religious commitment of Kester, Williams, and Rodgers proved a potent one.

"When they first started talking about union," confessed one white sharecropper, "I thought it was a new church." Indeed, STFU organizers quickly adopted a religious style of exhortation, language, and imagery in order to reach the sharecroppers in a familiar idiom. Exodus, the Old Testament story of deliverance from bondage in Egypt, appeared particularly appropriate, with the planters standing in for Pharaoh. Most telling was the singing that came to characterize an STFU gathering, which blended the tunes of familiar religious hymns with class-conscious pro-union lyrics, often supplied by the union's unofficial bard, John Handcox, a black sharecropper and organizer from St. Francis County. "Be Consolated" provides a typical example of Handcox's craft:

Often you go to the shelf and get your Bible,
And sit down and begin to read;
And you find therein where God will punish
The rich man for every unjust deed.

Be of good cheer; be patient; be faithful,
And help the union to grow strong.
And if at any time you become the least discouraged,
Revive yourself by singing the good old union song.[29]

"The union is not merely an industrial organization," wrote one of its sympathizers in *The New Republic* magazine, "it is a religious organization as well." Not surprisingly, in 1935 nearly half of the union's fourteen-member executive council were preachers; "they are by their deeds social revolutionaries with a religious drive," concluded Kester.[30]

These preachers were men, of course, but women proved especially amenable to the union gospel. As an economic system, sharecropping depended heavily on a family economy that required women's labor as much as men's, but as traditional heads of households the menfolk were the ones who at first became official union members. Before long, however, drawn both by their own sense of economic hardship and their religious impulse and participation, women insisted on full-fledged union status for themselves. Many of them became organizers and activists within the STFU. At the local level, female STFU members provided a crucial skill: they were literate. Consequently, black and white women served as business secretaries for many STFU locals across rural Arkansas. For them as well, the collective experience of the union wedded to the rural grassroots church helped open the doors to an entirely new consciousness of self-emancipation.

In the Arkansas Delta in the 1930s Kester found in ordinary southerners a living example of the religious precepts and spiritual val-

A preacher (perhaps E. B. McKinney) holds his bible aloft in an open-air "church." This unidentified photograph illustrates the importance of religious oratory to the union's constituency. If a sermon spoke the biblical language of bondage and deliverance, the worldly social import of the message to the poor of the Arkansas Delta was unmistakable. As natural leaders of their community, African American preachers readily doubled as STFU organizers. Kester and other STFU activists adopted the prophetic rhetoric of the Old Testament to speak to sharecroppers in a familiar revolutionary language. STFU Papers, Southern Historical Collection, Library of the University of North Carolina at Chapel Hill.

ues that had stirred his political imagination for the past decade. In many ways, the STFU marked the logical culmination of all of his previous social activism. Its socialist politics and leadership appealed to his goal of radical social transformation; its interracialism put into practice an ethic Kester had long tried to encourage and live by; its non-violent tactics appealed to his pacifism; and his agrarianism found expression in the efforts of the literally "disinherited" southern landless class to make a better life for themselves on the land. All of these elements of Kester's faith in the STFU can be found in *Revolt among the Sharecroppers*. Less obvious, however, is the way the very project of building the union may have undermined them.

II

Kester worked primarily as a social activist, not an intellectual or a writer. In his devotion to the STFU he labored tirelessly to organize the union, consult with its leaders and rank and file, and plead its case before concerned clergy and liberals far removed from the plantation South. The fact that he chose to devote several months of his time to telling the story of the STFU in print suggests the depths of his commitment to the union's cause. But Kester wrote about an organization that, in his mind, might usher in the Kingdom of God; like all organizations, the STFU proved to be all too human.

Written during the summer and fall of 1935, *Revolt among the Sharecroppers* could only chronicle the remarkable history of the STFU's first year, during which Kester served the union as an active participant. Despite the intense repression visited on the union and its members, extensively documented by Kester, the book ends on an immensely hopeful note. The AAA had failed to respond to the sharecroppers' pleas, but this moved them to take matters into their own hands. Thus in the fall of 1935 the STFU made the transition from a protest organization to a union by calling a strike. This action, described by Kester at the close of his book, won the union a great deal of prestige within the labor movement and helped it grow by leaps and bounds. Between October 1935 and the end of the year the STFU chartered seventy-one new locals, and claimed to have twenty-five thousand members by 1936; the union also began to expand beyond its base in Arkansas, establishing locals in Oklahoma, Missouri, and Texas, though the bulk of its membership remained concentrated in the Arkansas Delta.[31]

Moreover, because of the terror the union faced, the sharecroppers soon won the attention of many powerful people in Washington, including Eleanor Roosevelt, Arkansas Senator Joe Robinson, and Wisconsin Senator Robert LaFollete. In January 1936, while his book was at the printer, Kester narrowly avoided being lynched himself. Speaking at a church in Crittenden County to protest the evictions of over one hundred people from a local plantation, Kester was kidnapped by a gang of armed men who broke into the meeting. Driven into the woods to be hanged, he managed to persuade his abductors that his murder would be investigated by FBI agents who would identify its perpetrators and bring them to justice. The would-be lynchers reconsidered and escorted

March at Marked Tree, Arkansas, February 1935, photo by Howard Kester. Through its affiliation with the Socialist International, the STFU commanded solidarity from abroad. This well-known march, organized by Kester and described by him in *Revolt among the Sharecroppers,* featured the participation of Jennie Lee and Zita Baker, prominent radical members of the British Labor Party, and Naomi Mitchison, author of an eyewitness account of the 1934 Socialist uprising in Vienna in the face of Nazism. Baker and Mitchison appear at the head of the march; Lee later addressed the crowd. STFU Papers, Southern Historical Collection, Library of the University of North Carolina at Chapel Hill.

Kester to the bridge at the Tennessee line with a warning not to return to Arkansas unless he wanted to be shot on sight.

This incident, and the many others like it, won the STFU a good deal of sympathy. By the time *Revolt among the Sharecroppers* appeared in March 1936, Senator Robinson, in an effort to avoid embarrassment for the Democratic Party, had exerted his immense influence in Arkansas to temporarily quell the worst repression. Senator LaFollete, at the behest of STFU Washington ally Gardner Jackson, agreed to investigate civil liberties violations against the STFU and the rest of the new industrial union movement; the hearings began the same week that Kester's work saw print but ironically never did consider the repression in the Delta. Back in Arkansas, the terror only increased the union's appeal to the sharecroppers. "We know the union is all right now, for the big bosses is trying to bust it up," one cropper informed Kester and Mitchell.[32]

Subsequent events, however, in which Kester continued to play an important part, revealed some of the union's inherent weaknesses. By the end of the decade three related internal conflicts left the STFU in disarray: tension over the issue of racial integration, bitter factional fighting between Communists and Socialists interested in organizing agricultural workers, and the fateful decision to join forces with the national movement to build industrial unions, the CIO.

The STFU and the CIO

Prior to 1935 the official face of organized labor in the United States was the American Federation of Labor (AFL), dominated by unions representing specific skilled crafts. The Great Depression and the subsequent government legitimation of trade unions with the 1935 National Labor Relations Act (the Wagner Act) stimulated an upsurge of labor organizing. Unable and unwilling to accommodate the new influx of eager unskilled recruits from the mass-production industries, the AFL soon found itself challenged by its breakaway rival, the Congress of Industrial Organizations (CIO). Unlike the AFL, the CIO unions opened their ranks to men and women, blacks and whites, immigrants and the native-born, skilled and unskilled, in the auto plants, rubber factories, slaughterhouses, coal mines, and steel mills of America.

When United Mine Workers President John L. Lewis resolved in 1936 to build this new national labor federation of unions organized by industry, rather than craft, he asked Communist Party (CP) member Donald Henderson to spearhead organizing efforts among workers in agriculture and affiliated industries. Henderson, a one-time economics instructor at Columbia University, had by the mid-1930s become the leading CP expert on agricultural laborers and edited the party's journal on such matters, *The Rural Worker*. Lewis, long hostile to Communists within the ranks of the miners' union, nevertheless recognized that party cadre made exceptionally dedicated and effective organizers. Despite the exclusion of agricultural laborers from the provisions of the Wagner Act, Henderson soon drew together workers in the fields and packing sheds of California's San Joaquin Valley, the fish canneries up and down the Pacific coast, the citrus groves of Florida, and the sugar-beet fields of the mountain west into UCAPAWA—the United Cannery, Agricultural, and Packing Workers of America. "We wanted to be a union," Mitchell recalled of the STFU many years later. "We wanted to be in the mainstream of organized labor." UCAPAWA seemed the natural home for the STFU, which sent delegates representing over thirty thousand members in seven states to the new union's founding convention in Denver in 1937.[33]

Donald Henderson once complained to Gardner Jackson that the STFU was "neither fish nor fowl." He and Mitchell feuded bitterly, as a matter of both politics and personality. "Two or three centuries ago," Mitchell remarked to Norman Thomas, "such people as Henderson were out hunting witches, and it is to be regretted that they are now engaged in attempting to organize farm laborers."[34] But perhaps Henderson had a point. For all its appeal as the avatar of the new industrial unionism, the CIO was designed for men and women who toiled in factories, mines, and packing sheds; earned wages; met in union halls; and paid dues. STFU members remained rooted on the land, farmed on shares, met in rural churches, and never had enough cash to pay the poll tax, let alone union dues. These were the factors that had shaped the union's basic character, its organizing strategy, its emphasis on social protest, its appeal to the conscience of the nation, and its religious fervor, but they sat quite uneasily within UCAPAWA. "Fish cannery workers on the West Coast do not have the same problem as cotton workers in the South," proclaimed the STFU in an open letter to its "friends" and allies explaining the reasons for breaking with UCAPAWA in 1939, after only eighteen months of affiliation. During this brief but troubled stint as part of UCAPAWA, the STFU never really could accommodate itself to its role as part of a CIO union. In retrospect, Mitchell admitted that "the ordinary trade unionist never understood the STFU. We were a mass movement, something like the civil rights movement thirty years later."[35]

The STFU and UCAPAWA quarreled over questions of procedure, finance, autonomy, and jurisdiction. Should locals correspond with the STFU office in Memphis or with UCAPAWA in Washington? How would the understaffed STFU and its uneducated local secretaries maintain the elaborate paperwork required by CIO unions and, more importantly, collect dues from its destitute members? The STFU was an organization with a weak structure, an amorphous "membership" far beyond those officially enrolled, and few resources; in the first six months of its CIO affiliation, the STFU could only come up with dues from one-third of its members, and the CIO provided very little in the way of support. And just who was the STFU responsible for organizing? Farmworkers in Arkansas and Oklahoma, or all across the South? Workers in cotton fields only, or all agricultural labor? Tenants and sharecroppers, or day laborers as well? In an effort to resolve these contentious issues, the STFU repeatedly pressed for "autonomy" within UCAPAWA, or barring that, direct affiliation with the CIO as a distinct organization. By 1938, Mitchell, Kester, and STFU president J. R. Butler sought clear jurisdiction over "all tenants, sharecroppers and farm laborers in the southern states" without interference from the UCAPAWA leadership.[36] This dispute over jurisdiction reprised a conflict that had long divided the Socialist leadership of the STFU from Communist Party organizers in the rural South, who had in the early 1930s built a parallel farm-labor union in Alabama, the Share Croppers' Union (SCU).

The STFU and the Communists

During the Popular Front years of the mid-1930s, the Communist Party sought to broaden its base by working closely with other left-wing organizations as well as liberal groups. Taking advantage of this brief moment,

Socialists and Communists forged a tactical alliance which did not entirely wash away the bitter enmity they had developed over the preceding period, when the CP had castigated Socialists as "social fascists." Like many Socialists in the left wing of the party, Kester regarded the CP with a mixture of admiration and skepticism. In the South especially, the Communists, many of whom proved courageous organizers in the face of violent repression, seemed to be fighting for the same things as Kester and his SP comrades: an end to lynching, along with civil rights for African Americans, economic rights for sharecroppers, labor unions for miners and textile workers, and an end to the iniquitous effects of the "fascist" AAA. Privately, however, as the Popular Front alliance crystallized in late 1934, Kester confided to a friend that "while I have been extremely sympathetic with the Communist position and have been very close to the organizers in this area I am becoming increasingly skeptical of the validity of their position and of their political tactics," which he believed invited racial violence.[37] The CP's most famous southern campaign during the 1930s was its efforts to free the "Scottsboro Boys," nine young black men falsely accused of raping two white women in Alabama, a case which Kester himself had investigated for the FOR in 1931. Even then he had felt that the CP's role in the Scottsboro affair was counterproductive and brought an already tense racial situation to the boiling point.

For its part, the CP greeted *Revolt among the Sharecroppers* with a hostile review in its newspaper, *The Daily Worker*. Apparently the Communists were "quite irritated" with Kester for limiting mention of the Alabama SCU to the few words of solidarity offered in his preface. "I am sincere in my desire to work with any group down here," Kester informed his friend Gardner Jackson, "but it is nauseating as hell to be told over and over again that you don't have the 'right line'."[38] Despite his attraction to the CP, Kester was never entirely comfortable with the party's apparent fealty to policies devised in the Soviet Union or with what he saw as its end-justifies-the-means approach to human affairs. He also found it difficult to reconcile the party's strictly materialist view of human nature with the religious dimension of his own revolutionary politics.

The inauguration of the Popular Front opened the opportunity for collaboration, or perhaps even a merger, between the STFU and the Communist-led Alabama SCU. The SCU, organized in 1931 by an

African American Communist from Birmingham's steel mills, had, like the STFU, faced extensive repression when it attempted to recruit dispossessed tenants and croppers in the cotton belt. In contrast to the STFU, however, the SCU was an almost entirely black organization, in part because it preached the Communist doctrine of "self-determination" in the black majority counties of the "black belt." "An appeal to uneducated negro field hands on the basis of Negro rule will drive every white man away from them," Mitchell complained of this CP approach to organizing the South, with a certain degree of insensitivity.[39] Moreover, by 1936 the SCU followed a new Communist Party policy that farm-labor unions should be divided by tenure: one organization for day laborers and sharecroppers, who essentially constituted a rural proletariat, and another for small farm owners and tenants, who retained a modicum of independence from the plantation lords.

In one of its earliest programmatic statements, the STFU had declared itself open to "laborers, share-croppers, renters, or small landowners whose lands are worked by them selves."[40] This solid commitment to organizing people who worked the land, whatever the particulars of their relationship to capital, remained central to the STFU's philosophy, political program, and status as an organization throughout its history. Such inclusiveness proved irreconcilable with the SCU's program of division by land tenure, derided by Mitchell as "craft unionism in the cotton fields."[41] An internal STFU report considering cooperation with the SCU pointed to another alleged distinction between the two organizations: "the SCU is not a mass-movement native to the soil" but a creature of the CP.[42] In the end, distrustful of the Communists' motives, uneasy with their emphasis on black self-determination, unconvinced of their influence among sharecroppers, and completely at odds with their organizational strategy for farm labor, the STFU rejected cooperation with the SCU and the Communist Party.

Many of the differences between the SCU and the STFU resurfaced upon the latter union's affiliation with the Communist-led UCAPAWA. Claude Williams, a friend of Kester's from Vanderbilt and one of the most outspoken advocates of CIO affiliation, was expelled from the STFU in 1938, charged with plotting a Communist "capture" of the union. Even before the discovery of Williams's alleged treachery, Mitchell had declared, with more prescience than he knew, that he "would rather see every local of the STFU disbanded than to see it in [the Communists'] control."[43]

Williams also joined forces with E. B. McKinney, who had become disillusioned with the STFU's approach to interracialism. The black STFU vice-president tried to build a separate organization of African American sharecroppers and "divide the union racially," in the words of STFU President J. R. Butler. McKinney had helped organize one of the STFU's largest interracial locals. But he had at one time been a follower of black nationalist Marcus Garvey and continued to harbor doubts about the blessings of black participation in integrated organizations. He suspected, with some justice, that the STFU's African American members bore the brunt of repression while whites reaped the benefits of organization. McKinney, complained Ward Rodgers, who remained a staunch interracialist, "is race conscious first and class conscious second."[44] When the inevitable showdown with UCAPAWA came, both Williams and McKinney, along with their considerable numbers of followers, remained in the CIO union.

No comparison of the STFU's approach to organizing sharecroppers with that of the Communists would be complete without mention of the other compelling first-hand account of agricultural unionism during the 1930s, the oral memoirs of SCU member Ned Cobb, *All God's Dangers*. Cobb, known in his testament as Nate Shaw, was an African American farmer in the Alabama black belt. Disturbed by the evictions and repossessions that accompanied the depression in rural Alabama, fed up with the racism that made black economic progress and self-sufficiency next to impossible, and excited by the SCU's militant defense of black rights, he eagerly joined the Communist-led union in 1932.

In contrast to Kester's political manifesto, Cobb tells the story of agrarian radicalism from the point of view of an illiterate black sharecropper uninterested in the national politics of the New Deal or the exhortations of socialism. Instead of a utopian vision of collective human emancipation, his memoirs record the individual strivings of a black tenant farmer in the rural South. Moreover, Cobb's desire for economic autonomy is given voice in absolutely authentic oral testimony, told from memory to a third party long after the events it recalls. As a genre, nothing could be more distinct from Kester's pamphlet, narrated more from the perspective of an observer than a participant, despite its author's intimate involvement with the STFU's daily affairs. As a document, *All God's Dangers* stands in eloquent counterpoint to *Revolt among the Sharecroppers*, and they may fruitfully be read in tandem. Their differences reveal the distance between the political program of the left, whether Socialist or Communist, and the more prosaic aspirations of those they sought to organize. Read together, however, their commonalties suggest that despite their sectarian disputes, the SCU and the STFU planted themselves in the same soil, fertilized by a set of common grievances and hopes.

The Cooperative Vision of the STFU

It would be a mistake, however, to attribute the STFU's troubles entirely to the sectarian squabbles between Socialists and Communists, intense as they were. More significant was the larger question of developing a radical social vision tailored to the realities of life in the rural South. "The cotton fields of the South present a challenge to unionism," proclaimed STFU leaders with considerable understatement, upon announcing their split with UCAPAWA.[45] The delicate racial dynamics of the region, the peculiar and fluid economic arrangements of sharecropping, and the lack of any viable social institutions other than the church made indigenous leadership essential to the success of a sharecroppers' union. This, the STFU claimed, UCAPAWA had refused to countenance or provide, just as it had failed to understand the difficult task of organizing impoverished rural men and women unversed in industrial unionism. The South, "unique and fundamentally different from the rest of the nation," required an organizing technique appropriate to its regional peculiarities which only the STFU and its grassroots leadership knew how to provide. The union had organized a group of people deemed unorganizable for generations; it had melded agrarian socialism with fundamentalist and prophetic Christianity; it had demonstrated the possibilities of interracialism in a Jim Crow society. Most importantly, unlike UCAPAWA, which continued to advocate division by tenure, the STFU sought to address the common problems of tenants, sharecroppers, and laborers, whose status was frequently "interchangeable" and might vary from year to year, contract to contract. These men and women, the STFU understood, "live together . . . and work together on the same plantation. The people themselves do not recognize any division."[46]

Despite its Communist leadership, UCAPAWA was a trade union, not a revolutionary organization, and its main task was to improve the wages and working conditions of its constituency. But for the members of the STFU, "the union is to many of them their religion . . . [and] the only organization which these people have ever had." In language clearly inspired by Kester, in what must have been one of his last contributions as an active participant in the union, STFU leaders insisted in an open letter that their organization was "a successful combination of a union and a movement for social emancipation" and that, by its ignorance of life in the Arkansas Delta, UCAPAWA threatened to destroy this fundamental aspect of their project.[47]

"I have never looked upon our organization as a union for the simple reason that the necessary elements for building a union are not present" in the plantation South, Kester noted in an unpublished autobiographical sketch written in the late 1930s.[48] Indeed, it is hard to imagine an ordinary CIO union, even one led by radicals, endorsing the following principles, ratified at the STFU's 1937 annual convention (the last before affiliation with the CIO):

> All actual tillers of the soil [should] be guaranteed possession of the land, either as working farm families or cooperative associations of such farm families;
>
> The earth is the common heritage of all, and the use and occupancy of land shall constitute the sole title thereto;
>
> This organization is dedicated to the complete abolition of tenantry and wage slavery in all its forms, and to the establishment of a new order of society wherein all who are willing to work shall be given the full products of their toil.[49]

As is evident from *Revolt among the Sharecroppers* and the autobiographical sketch he wrote a few years later, for Kester the STFU embodied a deeply utopian vision of radical social transformation and spiritual awakening far beyond anything offered by the new industrial unionism. "[Even] before I started working among sharecroppers I envisioned a complete reordering of life in the rural South, the building of a new civilization of peace, abundance, beauty, and bread," Kester admitted. "I am interested, but I do not get very excited over the mere struggle for shorter hours and more pay," he continued. To restore "beauty in their lives," the landless of the South must be given the land and "freedom from drudgery and monotonous toil, from loneliness, emptiness, and despair" as well as "ignorance and poverty." Wedded to his socialist beliefs, Kester's Christian faith envisioned "men possessed of the knowledge that they are loved and wanted; that they have talents which are needed in building the world of tomorrow." For all its genuine egalitarian promise, the CIO appeared a rather poor vehicle for ushering in this "Kingdom without frontiers, a garden of Eden here in the South."[50] Kester yearned for nothing less than something akin to the "cooperative commonwealth" once dreamed of by the Debsian socialists, the Populists, and even their nineteenth-century forebears in the Knights of Labor.

Nowhere did Kester's vision of the restoration of humankind's organic relation to the land and the community solidarity of village life find truer expression than in the STFU's program of a cooperative agrarian society, described in some detail in the final chapter of *Revolt among the Sharecroppers*. The "new type of rural culture" imagined by Kester fell somewhere between the forced mass collectivization of agriculture then going on in Stalin's Russia and the wholly inadequate programs of the New Deal, which continued to emphasize and encourage single-family farm ownership. Unlike the disastrous Soviet experiment, Kester advocated voluntary collectivization and foresaw the ownership of some private property as compatible with cooperative farming. In contrast to some of the New Dealers, the Nashville Agrarians, and, it should be said, many of the sharecroppers themselves, he realized that by the 1930s the long-denied promise of "forty acres and a mule," dating back to Radical Reconstruction, was impractical. Instead of a return to the small family homestead, Kester and the STFU embraced some of the central features of modern agriculture: large-scale units, mechanization, diversification, and aid from scientific experts. They hoped to end dependence on a commodity like cotton and instead cultivate crops that could supplement the deficient sharecropper diet and broaden the southern tenant's market. Most of all, they intended to humanize and democratize agricultural efficiency so that those who worked the land would reap the benefits of the surplus it produced; the STFU would replace the social organization of the plantation with that of the cooperative farm.

In Kester's view, the New Deal, based as it was on restoring prices through a policy of scarcity, benefited the powerful at the expense of the powerless. The alternative, he and his fellow Socialists believed, was to harness the productive capacity of the modern economy for the fulfillment of human needs rather than for profit. "Capitalism and its price system were born of scarcity," claimed the venerable Norman Thomas in *The Plight of the Share-Cropper*. "We shall never overcome the economy of scarcity and truly accept the economy of abundance until we think . . . of what collectively we might have in terms of our needs and our resources" he concluded.[51] As Kester and other STFU activists accurately perceived, even at their best the New Dealers and the Democratic Party proved unwilling to take such a radical step other than on the most paltry scale.

Carry It On: The Demise and Legacy of the STFU

"Bad as the system of sharecropping is," warned STFU Acting Secretary J. R. Butler in 1935, "the break-up of that system will be infinitely worse."[52]

When Cross County planter C. H. Dibble agreed to bargain with the STFU in 1936, the local bank immediately threatened to close on his mortgage and no one would gin his cotton. Faced with this community pressure, Dibble adopted a familiar anti-union tactic: he evicted the members of the union, pictured here, from his land. Some of these same families formed the nucleus of the STFU's cooperative rural community across the river in Mississippi, the Delta Cooperative Farm. STFU Papers, Southern Historical Collection, Library of the University of North Carolina at Chapel Hill.

In truth, the very forces that gave life to the STFU also wrote its epitaph, for by 1935 the sharecropping system had become part "of an unreclaimable past," in the words of the sociologists who wrote *The Collapse of Cotton Tenancy.*[53] The AAA's crop-reduction program, by stimulating evictions, providing planters with cash for mechanization, and encouraging a shift to wage labor, hastened the dissolution of the old plantation order in the South. Yet as Kester acknowledged, social, economic, and technological forces far beyond the control of ordinary rural southerners or their advocates had already eroded the sharecropper's way of life and begun to usher in the era of capital-intensive agriculture. This meant, however, that the STFU sought to organize a fugitive class at the very moment of their uprooting from the land; indeed, between 1930 and 1940, nearly 200,000 African American tenants abandoned the soil, as did 150,000 whites.[54] In the long run, these families entered the stream of migrants to the North and to the West. Black Americans in particular left the South in the millions for life in Los Angeles, Chicago, Detroit, Cleveland, Philadelphia, and other cities in the decades following the depression. In the process they helped transform the nature of race relations, urban politics, and the struggle for civil rights. Although this exodus marked the opening up of unprecedented opportunities for both African American and white rural people, something vital was lost as well. The vanishing agrarian values so cherished by someone like Kester proved poorly adapted to modern urban life.

The framers of the New Deal attempted to respond to this rural upheaval, first with the Resettlement Administration, and then with the FSA, established as part of the Bankhead-Jones Farm Tenancy Act of 1937. Inspired in part by the political pressure exerted by the STFU and other groups concerned with alleviating rural poverty, the FSA provided low-interest loans to tenant farmers seeking to purchase their own land and "rehabilitation" loans to impoverished sharecroppers who needed a leg up the agricultural ladder. In an effort to resettle displaced agricultural workers, the FSA even experimented with rural cooperative communities, not unlike those envisioned by the STFU. This marked one of the New Deal's most striking departures from competitive individualism. Nevertheless, Norman Thomas regarded the FSA programs as "simply a gesture . . . to divert attention from some of the results of the AAA."[55]

Even if we accept its good intentions, the FSA, like the Resettlement

Administration before it, sought to bail a sea of misery with a tiny bucket. By 1940, less than 1 percent of the tenant families in the South had received purchase or rehabilitation loans from the government. In 1939, for instance, 150,000 farm families applied for FSA loans; 7,000 were granted. Of the 192,000 displaced rural African American families during the 1930s, only 2,000 received FSA purchase loans, and another 1,400 resettled in FSA cooperative communities. Even so, this program proved too radical for the southern wing of the Democratic party, and by 1944 they joined with conservative Republicans to kill the FSA.[56]

The closing of the New Deal era also marked the demise of the STFU. The sparks struck by the union could only burn bright in the oxygen provided by the atmosphere of the decade itself. The labor upsurge, the climate of sympathetic liberalism, and the faith that government action could and should alleviate the lives of the destitute made the times ripe for a sharecroppers' union. But the contradictory quality of such an organization became clear as the labor movement consolidated itself, leaving little room for the STFU, which as Kester had always understood, represented more than a union. By 1939, *Revolt among the Sharecroppers* had gone out of print, and the STFU emerged from its break with UCAPAWA with only forty of three hundred locals intact.[57] Disillusioned by the CIO debacle and the political bickering that had accompanied it, Kester himself had returned to religious activism and become increasingly involved with the Fellowship of Southern Churchmen, a radical Christian alliance he had helped found several years before.

Much of the organized concern that made the 1930s such a promising time for social activists evaporated as the country turned to war and recovery during the 1940s. The wartime economy drew many of the rural workers who formed the STFU's constituency into the booming factories and shipyards, and thus into the CIO; indeed, during World War II the STFU served its dwindling membership primarily as a labor recruiter rather than as a union or a social movement. Renamed the National Farm Labor Union (NFLU) after the war, the union affiliated with the AFL. Although in its new incarnation the NFLU took on the struggles of farmworkers in California and Louisiana, among other places, it never regained its identity as a radical social movement. Meanwhile, the rise of the Cold War and its attendant domestic hostility to anything that smacked of "communism" stymied the most progressive wing of New Deal reform and hamstrung much of the impetus for bringing radical social change to America that was its adjunct. By the end of the 1940s, southern Democrats and their allies easily equated unionism, opposition to segregation, and any effort to challenge the status quo with the threat of Soviet Communism and outright subversion.

Nevertheless, the ground did not lie fallow between the dampening of 1930s social activism and the awakening of the 1960s, as is commonly assumed. A small group of dedicated southerners sought to break the hold of the "Dixiecrats" on the Democratic Party and continued to press for labor unions, abolition of the poll tax, antipoverty programs, and an end to segregation. Between the CIO's failed postwar attempt to organize the South and the Montgomery Bus Boycott of 1956, which inaugurated the modern civil rights movement, stood a scant ten years. Thus, if the STFU as an organization withered on the vine, its spirit easily flowered again only two decades later with a renewed effort to democratize the South using the tactics of non-violence. The STFU pioneered a form of what the New Left of the 1960s came to call "participatory democracy," the ethic that "the individual share in those social decisions determining the quality and direction of his life," in the words of the famous 1962 Port Huron Statement. (Students for a Democratic Society [SDS], the progenitor of this manifesto, was the youth wing of the League for Industrial Democracy and thus a direct descendant of Norman Thomas's Socialist Party.) The socialist and religious activists epitomized by Kester, who laid their lives on the line in the cause of the sharecroppers during the 1930s, found their reflection in the young people of the 1960s who joined hands with the civil rights movement in freedom rides, sit-ins, voter-registration drives, and organizing projects.

Across the Mississippi River and only a few hundred miles south of where the STFU had sunk its roots, the Student Nonviolent Coordinating Committee planted new seeds in 1961. In this organization young whites and blacks sought to create a "beloved community" that would help poor black people in the Mississippi Delta emancipate themselves from poverty and dependence. SNCC activists made a concerted effort to challenge the unbroken power of Delta plantation owners, still made rich by federal agricultural subsidies and still represented by the southern wing of the Democratic party. (Unlike the STFU before it, however, SNCC had little success organizing poor southern whites). Finally, the Christian socialism and pacifism of Kester and his fellow religious radicals of the 1930s were revived in the thought of Martin Luther King Jr. Like these predecessors (and drawing on many of the same sources of inspiration), King fused his hope for a more egalitarian economic order with a spiritual faith in the possibility of human redemption and brotherhood, especially across the barrier of race. King too was a southerner, and he believed as Kester had that only a non-violent interracial movement of the poor could move the nation in this direction.

The STFU spirit and legacy might be detected somewhere else as well. By the 1960s the center of American commercial agriculture had shifted decisively from the South to the "factories in the field" of California, where huge numbers of itinerant Mexican American and Mexican workers labored for giant agribusiness corporations. Twenty-five years after the New Deal, these farmworkers still remained outside the protections of the liberal welfare state, including the Wagner Act and the Fair Labor Standards Act of 1937, which mandated minimum wages. The United Farm Workers, led by Cesar Chavez, sought to organize these men and women into a union. Under Chavez, the UFW merged the quest for collective bargaining rights with non-violent social protest.

As had STFU activists before them, UFW partisans understood that agricultural unionism, written out of labor law, must rely on public sympathy and the active support of radicals, liberals, and concerned clergy to succeed. Thus Chavez pressured the California growers with the power of the public consumer boycott, most famously the campaign against table grapes. Social activists from SDS, SNCC, and liberal religious denominations flocked to the UFW's picket lines and meeting halls and went to jail on the union's behalf. And, like the interracial STFU, whose cause was linked to the long-deferred emancipation of African Americans, the UFW injected a healthy dose of civil rights consciousness into its struggle to liberate the Mexican Americans who harvested America's fruits and vegetables. One of the UFW's contemporary chroniclers, environmentalist Peter Matthiessen, perceived that the farmworkers' movement of the 1960s aimed at nothing less than "an order of things that will have humanity as its purpose and the economy as its tool, thus reversing the present order of the system."[58] Buck Kester could not have said it better.

In the last pages of *Revolt among the Sharecroppers,* Kester recounts his contemplation of a radish gracing his dinner plate in a Memphis hotel. Across six decades his meditation on the human exploitation incarnated in that small product of labor on the land still resonates. Kester's modest contribution to the flood of social-protest literature of the 1930s remains a pertinent reminder that the enormous material bounty enjoyed by many Americans has come with a high cost—then, and now. As H. L. Mitchell said, the story Kester tells "is powerful because it is true."[59]

Alex Lichtenstein

NOTES

Notes are provided only for directly quoted material and numerical figures. For complete citations and a bibliography, see the Works Consulted and Further Readings following the notes.

1. Howard Kester to Norman Thomas, Aug. 18, 1937, Southern Tenant Farmers' Union (STFU) Papers, microfilm edition, reel 59.
2. Kester, autobiographical sketch, c. 1937, p.11, STFU Papers, reel 60.
3. Martin, *Howard Kester and the Struggle for Social Justice,* 2; Kester, autobiographical sketch, 13.
4. Kester, autobiographical sketch, 7–8.
5. Dunbar, *Against the Grain,* 29.
6. Martin, *Howard Kester and the Struggle for Social Justice,* 60.
7. Ibid., 81.
8. Green, *Grass-Roots Socialism,* xi.
9. Dunbar, *Against the Grain,* 91.
10. Donald Holley, "The Plantation Heritage: Agriculture in the Arkansas Delta," in Whayne and Gatewood, eds., *The Arkansas Delta,* 258.
11. Ibid.
12. Grubbs, *Cry from the Cotton,* 10.
13. Wolters, *Negroes and the Great Depression,* ix, 9, 84.
14. Ward H. Rodgers, "Sharecroppers Drop the Color Line," *The Crisis* 42 (June 1935):168.
15. Irons, *The New Deal Lawyers,* 160.
16. Daniel, *Breaking the Land,* 98–99.
17. Mitchell and Butler, "The Cropper Learns His Fate," *The Nation* 141 (Sept. 18, 1935):328.
18. Howard Kester, "The Struggle Against Peonage," address to 26th Annual Conference of the NAACP, June 26, 1935, pp. 7, 11, STFU Papers, reel 59.

19. Johnson, Embree, and Alexander, *The Collapse of Cotton Tenancy,* 10; Rodgers, "Sharecroppers Drop the Color Line," 168; Martin, *Howard Kester and the Struggle for Social Justice,* 5.
20. Raper, *Preface to Peasantry,* 5.
21. Thomas J. Watson, "The Negro Question in the South," 371.
22. "History of the STFU," typescript, c. 1936, p. 2, STFU Papers, reel 3.
23. Grubbs, *Cry from the Cotton,* 67.
24. Dunbar, *Against the Grain,* 285.
25. Ibid., 107.
26. Rodgers, "Sharecroppers Drop the Color Line," 169; Lucien Koch, "The War in Arkansas," *The New Republic,* Mar. 27, 1935, 184.
27. J. E. Clayton to H. L. Mitchell, Dec. 28, 1937, STFU Papers, reel 5; "Report—Share-croppers Union of Alabama," c. 1935, p. 4, STFU Papers, reel 1.
28. Howard Kester to Richard Gregg, July 28, 1937, STFU Papers, reel 59; Miller, *Working Lives,* 134.
29. Workers Defense League, *The Disinherited Speak,* 31.
30. Grubbs, *Cry from the Cotton,* 64; Koch, "The War in Arkansas," 184; Dunbar, *Against the Grain,* 110.
31. Dunbar, *Against the Grain,* 108.
32. Mitchell and Kester, "Share-Cropper Misery and Hope," *The Nation* 142 (Feb. 12, 1936):184.
33. Miller, *Working Lives,* 140.
34. Mitchell, *Mean Things Happening,* 164; H. L. Mitchell to Norman Thomas, Nov. 17, 1936, STFU Papers, reel 3.
35. "An Open Letter to the Friends of the Southern Tenant Farmers' Union," Oct. 1, 1939, p. 3, STFU Papers, reel 13; Miller, *Working Lives,* 140.
36. Minutes of STFU National Executive Committee Meeting, Apr. 25, 1938, STFU Papers, reel 8.
37. Dunbar, *Against the Grain,* 103.
38. Howard Kester to Gardner Jackson, Apr. 9, 1936, STFU Papers, reel 59.
39. Mitchell to Howard Kester, Feb. 7, 1938, STFU Papers, reel 59.
40. "A Program for Action", Nov. 1, 1934, p. 1, STFU Papers, reel 1.
41. Kelley, *Hammer and Hoe,* 170.
42. "Report—Share-croppers Union of Alabama," 2.
43. Mitchell to Kester, Feb. 7, 1938.
44. Butler to Norman Thomas, Aug. 23, 1938, STFU Papers, reel 8; Ward Rodgers to Walter White, Mar. 3, 1938, STFU Papers, reel 59.
45. "An Open Letter to the Friends of the Southern Tenant Farmers' Union," 3.
46. "An Open Letter to the American Labor Movement and the Friends of the Southern Tenant Farmers' Union," May 21, 1938, pp. 2–3, STFU Papers, reel 8.
47. Ibid., 4.
48. Kester, autobiographical sketch, 11.

49. Dunbar, *Against the Grain,* 154.
50. Kester, autobiographical sketch, 11–12.
51. Thomas, *Plight of the Share-Cropper,* 18.
52. Butler to Lawrence Westbrook, May 15, 1935, STFU Papers, reel 1.
53. Johnson, Embree, and Alexander, *The Collapse of Cotton Tenancy,* 35.
54. Wolters, *Negroes and the Great Depression,* 33.
55. Ibid., 66.
56. Ibid., 66, 79.
57. Mitchell, *Mean Things Happening,* 182.
58. Matthiessen, *Sal Si Puedes,* 30.
59. Mitchell to All Locals, Apr. 8, 1936, STFU Papers, reel 2.

WORKS CONSULTED AND FURTHER READINGS

Primary Sources

Asch, Nathan. "Marked Tree, Arkansas." *The New Republic,* June 10, 1936, 119–21.

Johnson, Charles S. *Shadow of the Plantation.* Chicago: Univ. of Chicago Press, 1934.

Johnson, Charles S., Edwin R. Embree, and W. W. Alexander. *The Collapse of Cotton Tenancy.* Chapel Hill: Univ. of North Carolina Press, 1935.

Koch, Lucien. "The War in Arkansas." *The New Republic,* Mar. 27, 1935, 182–84.

Mitchell, H. L., and J. R. Butler, "The Cropper Learns His Fate." *The Nation* 141 (Sept. 18, 1935):328–29.

Mitchell, H. L., and Howard Kester, "Share-Cropper Misery and Hope." *The Nation* 142 (Feb. 12, 1936):184.

Raper, Arthur F. *Preface to Peasantry: A Tale of Two Black Belt Counties.* Chapel Hill: Univ. of North Carolina Press, 1936.

Raper, Arthur F. *The Tragedy of Lynching.* Chapel Hill: Univ. of North Carolina Press, 1933.

Raper, Arthur F., and Ira DeA. Reid, *Sharecroppers All.* Chapel Hill: Univ. of North Carolina Press, 1941.

Rodgers, Ward H. "Sharecroppers Drop the Color Line." *The Crisis* 42 (June 1935):168–69, 178.

Southern Tenant Farmers' Union. Papers. Southern Historical Collection. Wilson Library, Univ. of North Carolina, Chapel Hill. This collection is available on microfilm at many libraries or can be borrowed on interlibrary loan. Reels 1–14 contain material from the 1930s; reels 59–60 consist of relevant selections from the Howard Kester Papers, also at the Southern Historical Collection.

Thomas, Norman. *The Plight of the Share-Cropper.* New York: League for Industrial Democracy, 1934.

Watson, Thomas J. "The Negro Question in the South." *Arena* 6 (Oct. 1892). Rpt. in Norman Pollack, *The Populist Mind.* Indianapolis: Bobbs-Merrill, 1960.

Workers Defense League. *The Disinherited Speak: Letters From Sharecroppers.* New York: Workers Defense League, 1937.

Secondary Sources

Ayers, Edward L. *The Promise of the New South: Life After Reconstruction.* New York: Oxford Univ. Press, 1992. Comprehensive survey of southern history during era of Populism, segregation, disfranchisement, lynchings, and increase in tenancy.

Cobb, James C. *The Most Southern Place on Earth: The Mississippi Delta and the Roots of Regional Identity*. New York: Oxford Univ. Press, 1992. A social and cultural history of the plantation economy in a region just across the river from and quite similar to the Arkansas Delta.

Daniel, Pete. *Breaking the Land: The Transformation of Cotton, Tobacco, and Rice Cultures since 1880.* Urbana: Univ. of Illinois Press, 1985. Agricultural history of the modern South; has a very useful chapter on the AAA's cotton program and its effects.

Dunbar, Anthony P. *Against the Grain: Southern Radicals and Prophets, 1929–1959.* Charlottesville: Univ. of Virginia Press, 1981. The story of the socialist and religious left in the South during these years; focuses on James Dombrowski, Myles Horton, Howard Kester, Ward Rodgers, Don West, and Claude Williams.

Egerton, John. *Speak Now Against the Day: The Generation Before the Civil Rights Movement in the South*. New York: Knopf, 1994; paper ed., Chapel Hill: Univ. of North Carolina Press, 1994. Encyclopedic compendium of southern movements for social and racial change from the 1930s through the 1950s.

Green, James R. *Grass-Roots Socialism: Radical Movements in the Southwest, 1895–1943.* Baton Rouge: Louisiana State Univ. Press, 1978. Social history of agrarian and backwoods socialist predecessors of the STFU in Arkansas, Louisiana, Oklahoma, and Texas.

Grubbs, Donald H. *Cry from the Cotton: The Southern Tenant Farmers' Union and the New Deal.* Chapel Hill: Univ. of North Carolina Press, 1971. Remarkably, still the only monograph on the history of the STFU. Especially good on the relationship between the union and government programs.

Honey, Michael K. *Southern Labor and Black Civil Rights: Organizing Memphis Workers.* Urbana: Univ. of Illinois Press, 1993. Important history of interracial organizing by the CIO, Socialists, and the Communist Party in Memphis during the 1930s and 1940s.

Irons, Peter H. *The New Deal Lawyers.* Princeton: Princeton Univ. Press, 1982; paper ed., 1993. Includes detailed account of the factional battles inside the AAA over the cotton program.

Jones, Jacqueline. *The Dispossessed: America's Underclasses from the Civil War to the Present.* New York: Basic Books, 1992. Sweeping history of the southern rural poor and their struggle for a better life.

Kelley, Robin D. G. *Hammer and Hoe: Alabama Communists During the Great Depression.* Chapel Hill: Univ. of North Carolina Press, 1990. Important study of this segment of the southern left; includes material on the Alabama Share Croppers' Union and the Communist Party's approach to problems of rural labor.

Kirby, Jack Temple. *Rural Worlds Lost: The American South, 1920–1960.* Baton Rouge: Louisiana State Univ. Press, 1987. Impact of modernity on the rural south during the twentieth century. Includes a thorough account of the STFU.

Lemann, Nicholas. *The Promised Land: The Great Black Migration and How It Changed America.* New York: Knopf, 1991. Anecdotal history about migration of sharecroppers from the Mississippi Delta to Chicago in the decades after the Great Depression.

Martin, Robert F. *Howard Kester and the Struggle for Social Justice in the South, 1904–1977.* Charlottesville: Univ. of Virginia, 1991. Fine biography of Kester, and the only single-volume treatment of his life and times. Emphasizes his religious background and thought.

Matthiessen, Peter. *Sal Si Puedes: Cesar Chavez and the New American Revolution.* New York: Dell Publishing, 1969. A sympathetic first-hand account of the United Farm Workers union in California during the mid-1960s.

Miller, Marc S., ed. *Working Lives:* The Southern Exposure *History of Labor in the South.* New York: Pantheon, 1980. Contains an interview with former STFU white organizers J. R. Butler, Clay East, H. L. Mitchell, and African American organizer George Stith.

Mitchell, H. L. *Mean Things Happening in This Land: The Life and Times of H. L. Mitchell, Cofounder of the Southern Tenant Farmers' Union.* Montclair, N.J.: Allanheld, Osmun, and Co., 1979. Mitchell's memoirs, not always reliable, but filled with fascinating anecdotes. Includes a foreword by socialist Michael Harrington.

Mitchell, H. L. *Roll the Union On: A Pictorial History of the Southern Tenant Farmers' Union.* Chicago: Charles H. Kerr, 1987. Great pictures and examples of union songs; introduction by Orville Vernon Burton.

Roediger, David R., and Don Fitz, eds. *Within the Shell of the Old: Essays on Workers' Self-Organization.* Chicago: Charles H. Kerr, 1990. Contains an interview with H. L. Mitchell.

Shaw, Nate [Ned Cobb], with Theodore Rosengarten, comp. *All God's Dangers: The Life of Nate Shaw.* New York: Knopf, 1974. Oral memoirs of an Alabama tenant farmer active in the Communist-led Share Croppers' Union.

Whayne, Jeannie. *A New Plantation South: Land, Labor, and Federal Favor in Twentieth*

Century Arkansas. Charlottesville: Univ. of Virginia Press, 1996. A recent economic and social history of Poinsett County, birthplace of the STFU.

Whayne, Jeannie, and Willard B. Gatewood, eds. *The Arkansas Delta: Land of Paradox.* Fayetteville: Univ. of Arkansas Press, 1993. Essays on history and economy of the region.

Wolters, Raymond. *Negroes and the Great Depression: The Problem of Economic Recovery.* Westport, Conn.: Greenwood Publishing Co., 1970. Includes a still very useful treatment of "agricultural recovery and the Negro farmer."

Wright, Gavin. *Old South, New South: Revolutions in the Southern Economy Since the Civil War.* New York: Basic Books, 1986. Definitive economic history of the South from the end of Reconstruction through World War II.

REVOLT AMONG THE SHARECROPPERS

EDUCATION, U.S.A.-SCHOOLHOUSE INTO BARN
THEY RIDDLED HIS HOUSE WITH BULLETS
WHERE THE EVICTED GO
CABIN IN THE WILDERNESS
WHERE KOCH AND REED WERE MOBBED
ONCE WE HAD A CABIN

TERRORS -
PICKING MACHINE
THE UNION MARCHES
THE COTTON
REVOLT
AMONG THE
SHARE-
CROPPERS
BY
HOWARD KESTER
50¢
BEATEN,
JAILED,
EVICTED
YOUTH
BEFORE EVICTION
AFTER EVICTION

REVOLT among the SHARECROPPERS

HOWARD KESTER

". . . A comprehensive account of the economic, social and political forces which have contributed to the desperate plight of the tenants and laborers in the cotton fields of the South."

New York · COVICI · FRIEDE · Publishers

MANUFACTURED IN THE UNITED STATES OF AMERICA
BY J. J. LITTLE AND IVES COMPANY, NEW YORK
TYPOGRAPHY BY ROBERT JOSEPHY

DEDICATION

To the disinherited sharecroppers everywhere and to the members of the Southern Tenant Farmers' Union who have found a new voice, created a collective will, discovered a forgotten hope, brought to life new impulses, given meaning to ancient concepts, struggled heroically against tyranny and oppression, risked their all that all might share and share abundantly, fashioned a new faith, built a mighty union, forged a new dream for the oppressed and disinherited and brought forth a song from lips where once hung a lamentation, this volume is affectionately dedicated.

FOREWORD

Howard Kester is devoting his life to the oppressed classes, Negro and white, of the South. As secretary of the Committee on Economic and Racial Justice he has been set free to work in the South wherever he can be most useful. Some of his time has been spent in inter-racial education and some of it has been devoted to work among the college youth of the South. But his most effective and strategic work has been given to the cause of the southern sharecroppers, who are probably the most exploited group of workers in any part of the western world.

In this pamphlet Mr. Kester gives a comprehensive account of the economic, social and political forces which have contributed to the desperate plight of the tenants and laborers in the cotton fields of the South. He proves the tragic inadequacy of the measures taken by the present administration for the improvement of agriculture. There is no more striking irony in modern politics than the fact that the provisions of the Agricultural Adjustment Administration, designed to alleviate the condition of the American farmer, should have aggravated the lot of the poorest of our farmers, the southern sharecroppers. He describes the achievements in organization of these poor farmers and emphasizes the necessity of such organization as a means of their ultimate emancipation. The organization of southern farmers, in which Mr. Kester is playing an important part, is incidentally an omen to be hailed for more reasons than one. It is achieving unity and coöperation between white and colored workers for the first time in American history. For that reason it holds out great promise and for the same reason it will face determined and ruthless opposition.

Since the invention of the cotton picking machine will undoubtedly tend to reduce the southern farm laborers to an even more abject state of poverty than that in which they find themselves today, it is safe to regard the field in which Mr. Kester works as the most crucial, as it is the most tragic in the whole social economy of America.

I hope, therefore, that this interesting and instructive analysis of King Cotton and his slaves by Mr. Kester will gain a wide circulation.

Reinhold Niebuhr

PREFACE

The purpose of this small volume is three-fold. First, to describe a general condition; secondly, to set down the labors of a particular organization working in the midst of these conditions; and thirdly, to suggest a way out.

The author chose the Southern Tenant Farmers' Union as the vehicle through which to express the revolt among the sharecroppers because of its extraordinary achievements and his intimate knowledge of its work. He is not unmindful of the magnificent contributions which the Sharecroppers' Union of Alabama has made toward securing better conditions for the cotton field workers, providing them with new incentives and a greater hope in the future and in making America conscious of the forgotten men of the cotton country. Their achievements are of great significance and the fact that the Sharecroppers' Union of Alabama is not specifically referred to in this volume is not to be construed as lack of appreciation by the author.

I have attempted to tell the story as I have seen and felt it, and to relate what I know to be true. All of the incidents herein related are real. The prologue is a true account of one of many similar experiences in which the author personally participated.

I wish to acknowledge my indebtedness to the several authors and publishers who have permitted me to quote from their works.

I owe special thanks to Mr. Norman Thomas for planting the idea of this volume in my mind and for much needed encouragement and advice; to Dr. Reinhold Niebuhr and Miss Elisabeth Gilman, both of whom were good enough to read the manuscript and who helped so much by their interest and encouragement and provided me and my family with our daily bread; to Mr. Gardner Jackson, Mr. Walter White, Dr. Broadus Mitchell, Dr. Ralph Harlow, Dr. A. D. Beittel, Rev. John Knox, Dr. William Amberson, Miss Dorothy Detzer, Miss Frances Williams, Miss Beth Torrey, Mrs. Charles Johnson, Mr. C. T. Carpenter and Mr. Clay East for their valuable suggestions and help. I shall forever be indebted to my colleague and friend, H. L. Mitchell, for insights into the mind of landlord and cropper and to Mrs. Mitchell for her devotion to all of us during the days when the union was struggling for its life. To the thousands of my friends in the Sharecroppers' Crusade in the South who are the real authors of these pages, I give my heartfelt thanks for the many lessons in courage, loyalty and justice they have given me.

I owe a special debt of gratitude to my wife, Alice, who helped in so many ways and without whose care and understanding this work would never have been done.

Howard Kester

Nashville, Tennessee,
November 29, 1935.

CONTENTS

I. MANHUNT

Just as the April sun showed its red face over the horizon, an automobile bearing three men slipped out of Memphis across the bridge connecting Tennessee and Arkansas and headed into the sharecropper country of northeastern Arkansas. The three men rode with mixed feelings of fear and jubilation. John Alden and his companions knew that if they were discovered by the planters and their retainers, there would be trouble. But trouble or no trouble there was work of a serious nature to be done. The men were going into Arkansas on urgent business for the Southern Tenant Farmers' Union; John Alden was going to help them in their work but he was also going to see his wife and children from whom he had been separated for many weeks. John Alden was returning to the plantation from which he had been driven by aroused planters and deputies who had become fearful of his union activities. The men rode in silence, watching passing automobiles and the workers in the fields.

John Alden said never a word—he was all eyes, seeing everything, missing nothing. He was full of memories too. As the driver of the car picked his way over the plantation roads, avoiding main highways, John Alden's tongue was loosened. He was in familiar country. In a small brick house lived the riding boss who gave him his eviction papers and threatened to kill him if he did not stop "organizing my niggers into that god-damn union." Across the wide acres he pointed to a brown smear where lived the president of their local and farther down a worn plantation road was the shack of the white sharecropper who had sheltered and protected him from the riding boss and his men. He was deeply moved as he talked. The words came from between sternly set white teeth and muscled jaws that were almost rigid.

"We were having a union meeting on that island there," he said, pointing to a small island in the middle of the St. Francis River. "After the meeting was over I decided to go across the river and see if I couldn't talk over some union business with some of our brothers down on Sixteen. After I had crossed the river, I met one of the riding bosses who asked me my name and where I was going. I told him that I was going to see some friends.

" 'You ain't got no business over here. You are just one of them damned union organizers who's been stirring up all our people. Let me see what is in that satchel!' "

Alden protested that the satchel contained only some Sunday School literature and a few song books.

"We've been hunting for you for six months and now I've got you. I'm going to arrest you but I'll need some help." The riding boss turned away to get help and Alden started toward the home of a white sharecropper.

As he reached the home of his brother sharecropper a car-load of heavily armed men drove into the yard. Alden was dragged from the porch out into the yard.

"Here's that union organizer we've been trying to catch so long!"

Guns were thrown in Alden's face. He was searched and the contents of his satchel dumped on the ground. The riding boss consulted his men. They would give Alden ten minutes to get off the plantation. If he did not make it in ten minutes they would shoot him. Alden asked for twenty minutes. The plantation, he said, was a big place. Someone said, "Give him a chance," and the time was extended to twenty minutes. The plantation stretched for miles along the banks of the St. Francis River. Down the banks of the river he plunged through marsh, water and woods. In twenty minutes he had lost himself in the back washes of the river. The human hunters came but did not find him. He lay in the weeds until darkness enveloped the land.

As night came on John Alden began to think of his wife and children. They would be uneasy about him, wondering what had happened. He thought of going back to say goodbye but he knew there was not one chance in a thousand for him to slip back to his cabin to see his wife and grab a few pieces of clothing. The riding bosses, planters and deputies would be on the lookout for him. Every road would be watched and covered. By this time word of his arrest and escape would have traveled throughout the countryside. They would be looking for this union organizer to shoot him at the first opportunity. There was but one thing to do. He would go to Memphis where H. L. Mitchell, the secretary of the Southern Tenant Farmers' Union, lived and there he would find protection and assistance. He faced Memphis.

Late that night he softly knocked on the door of a friendly Negro's cabin. They were looking for him. They thought that he would show up when night came and it was safe to travel. But he could not stay there for "the law" was watching every union man's home. He started walking. That night he spent beneath a bridge. As he lay on the naked earth his mind wandered back to the plantation. Trouble lay heavily upon him. He was cold and hungry. The ground beneath the bridge was as hard as the concrete above but about as comfortable as his corn shuck mattress in his cabin on the plantation. He had slept on worse than this. The soggy ground chilled him and he thought how good a cup of coffee would taste but he could not remember when he had had any coffee. As cold as he was, he was afraid to build a fire. He lay on his back looking up at the white concrete roof over him. A great hunger gnawed at his belly. He felt weak and faint. He was so hungry that he could eat some of that pork—red pork it was too—they had in the commissary, pork which had grown red from eating the carcasses of cows and mules which had died on the plantation.

A light flashed. White men were looking for something. They walked in silence and their guns gleamed in the moonlight. John Alden froze himself to the ground. The men came closer. Their flashlights covered the banks of the stream and their guns followed the light. John Alden merged with the brown earth. Near the bridge they stopped. Someone spoke softly. The men laughed. John Alden saw that they were not on a manhunt. They were hunting frogs.

The next day he watched from under the bridge. He traveled at night along the creek bottoms and deserted plantation roads. During the day he watched. For days he went without food. His boots were like so much lead about his feet. His clothes hung as though they were plastered to him. At times they dripped water. At West Memphis he met a union brother who had seen deputies looking for him. Carefully he made his way across the levee and to the river. On the Harahan Bridge he passed the line that divides Arkansas from Tennessee. He turned and looked back.

The Mississippi was rushing to the sea beneath him. Memories of long years of grinding toil in the cotton fields filled his mind. He thought of the twelve years of faithful service he had rendered the Swift Brothers on their vast plantation in Cross County. He saw a big boat on the Mississippi. It was loaded down with cotton, maybe some of the cotton he himself had picked and slaved for, going down the river to New Orleans and then to all parts of the world. He looked at the swirling waters of the river and a sense of frustration seized him. He had put in many hard years of labor, he had worked like a slave, half the time he was half-naked and more than half-starved and now he had nothing to show for it. He lived in a dilapidated shack. When it rained the water poured through the roof. In winter the wind tore through the holes in the walls. He could not recall when he had had a Sunday suit or an extra pair of overalls. The ones he had on were faded and covered with patches. He had worked day in and day out all through the year, yet work as hard as he might, each year found him in debt to the landlord. He had lived on cornbread, fat-back and molasses. He had forgotten there was anything else for men to eat. His life had not been his own; he lived as though he were a slave.

As he stood on the bridge and looked into the swirling, tossing water of the Mississippi beneath him, he mused. Well, wasn't he a slave? He had no rights, no privileges, nothing but poverty and grinding toil for his companions. All his life he had done nothing but slave and starve. If he objected to the way he was treated the riding boss beat him. Some men had been shot for questioning the will of a riding boss. When he and his fellow-sharecroppers organized the union to help one another they ran into a world of trouble. They were threatened by the riding bosses, evicted from their cabins and often beaten. The plantation manager had tried to evict him but he would not move. He had no place to go and besides he ought to have some rights. He had been faithful and done his work well. No one could say that he had been lazy and shiftless. Even Mr. Swift had said that he was a good farmer, reliable and honorable. He had worked, slaved, starved and he was not going to be run off the place unless there was a mighty good reason. Well, there was a good reason. The manager had told him what it was. It was the union. That was what was causing all of the trouble. The sharecroppers were getting together to end the slavery in the cotton fields. The planters were trying to break up the union.

As John Alden looked into the brown waters of the Mississippi to which the morning rays of the sun were giving a golden lustre, it all became clear. Yes, he had been a slave. He had been one among thousands. Now the sharecroppers were no longer willing to be slaves. Slaves begged for mercy but men demanded justice, and justice was all they asked. The planters were afraid of the union because it promised freedom to the enslaved sharecroppers. Now that the white and black slaves had stopped fighting one another and had joined together to struggle against their common enemy, the planter could no longer use the white man to beat down the black man or the black man to beat down the wages and living conditions of the white man. These once ancient enemies were together now. That made a difference—a world of difference. The Mississippi whirled on beneath him. John Alden's mind moved like the river.

The planters wre trying to break the union. Well, would they? Would they break the union and destroy the only hope the sharecroppers had of getting any justice? He thought of the men, women and children back on the plantation. They were slaves. Nearly everybody who worked for King Cotton was a slave. Now the slaves had a union—a real union composed of the white and black slaves of King Cotton. That was something new, something worth struggling for, yes, something worth dying for. The white and black slaves had stood up like men and fought shoulder to shoulder in the union. Now they were real men who understood the meaning of freedom and justice. The mighty river rolled beneath him and a great confidence and hope swelled in his heart. No, he was not going to turn his back on this thing. It was the one thing worth living for: it was the hope of his people—of all the enslaved sharecroppers everywhere in the South. He would wade in deeper than ever before. He had been secretary of his local on the plantation but now he would do even more. Yes, he would go into Memphis to the headquarters of the Southern Tenant Farmers' Union and see Brother Mitchell. He would struggle more than ever against the system which had enslaved millions of his brothers. He turned and walked across the bridge into Memphis where he found brothers and comrades.

Now he was back again at the home of the white sharecropper who had protected him from the bloodthirsty mob many weeks before. The three men talked with the union men whom they had come to see. A messenger was sent across the river to tell Mrs. Alden that her husband waited to see her. Hours passed and no one came across the river. No one spoke but everyone knew what the others were thinking. Perhaps a riding boss had discovered that John Alden was on the river and had raided his cabin. Maybe there had been a tragedy. Once in a while someone would relieve the strain by suggesting that John Alden's wife was in the field and could not leave without attracting attention, or that maybe she was going miles out of her way so as not to give her secret away. Finally, toward dusk a canoe slipped out of the bushes on the other side of the river and two women and a man came into view. Yes, they had come.

On the banks of the historic St. Francis which has been the silent observer of many tragic events in the lives of the varied people who have lived about it, John Alden met his wife. Death had been racing with both of them. They had felt his grasping fingers more than once during these last few months of terror and suspense. Now they had out-distanced him and life had joined them together once more.

John Alden met his wife. There were no tears, no sighs, no regrets. There were simple questions about health, food, clothing, crops, friends, asked and answered. He had brought the family some clothes which friends in the East had sent the union. The sun began to cast a purple shadow over the newly plowed land. The men bade their brothers goodbye. John Alden took his wife in his arms. "You and I have got to stand by the union, wife, it's our children's hope and ours."

The three men followed an unused plantation road and observed that darkness was their ally.

II. THE HERITAGE OF THE SHARECROPPER

"Since the Civil War, whether white or black, he had been the equivalent of slave labor."

—Coker

"The plantation system involves the most stark serfdom and exploitation that is left in the western world."

—Norman Thomas

Cotton to a greater degree than any other product of industry or agriculture rules the South. Indeed, one is almost justified in saying that the South lives by cotton. It determines the lives of more than twenty million people in the cotton producing states of the South. When cotton is high and the demand good the South is in the best of humor. When it is down and the demand poor one sees this fact registered in the faces of bankers, merchants, middlemen, landlords and planters. Cotton either makes the South happy, gay, and genial or harsh, bitter, and ugly. Sometimes the price of cotton is reflected on the sharecropper's face, but not often, for whether high or low his life continues just about the same.

The cotton belt extends something like 300 miles North and South and stretches for 1600 miles from the Carolinas to Texas. In this area cotton is produced for the most part by a semi-feudal mode of farming known as sharecropping. More than 8,000,000 individuals are capable of being classified as belonging to this category of farm tenure. "Although adding a billion dollars annually to the wealth of the world, the cotton farmers themselves are the most impoverished and backward of any large group of producers in America." [1]

It is generally recognized today that the southern sharecropper in the cotton fields is the most exploited agricultural worker in America. The conditions under which he lives have recently come to light to rudely awaken and shock the reading public of our entire nation. For the first time America is becoming conscious of the sharecropper who is the basic human element in the production of cotton.

Since the Civil War sharecropping as a system of producing cotton has been as fundamental to southern economy as banks and currency. When the Civil War closed the southern aristocracy turned to sharecropping as a means for continuing its existence. During these seventy years in which the system has been in operation, countless voices have been raised against the miserable conditions under which sharecroppers, tenant farmers and day laborers are forced to live and work, but they have been cries in a wilderness, unheard and unheeded, except by those wretched victims of King Cotton who were unable to help themselves.

Today the sharecropper is conscious of his condition and is attempting to make himself heard and felt through the united efforts of those in like circumstances. Thus through the Southern Tenant Farmers' Union the disinherited sharecroppers, tenant farmers and day laborers of both races are becoming increasingly articulate and gradually thrusting themselves into a nation's conscience. Through this united struggle they will either climb out of the hell of misery in which they have been floundering or they will be more completely submerged as America battles and blunders for a way out of the present chaos.

It is a commonplace to say that the South is different from the rest of America. It is so different, however, as to constitute an altogether different civilization. While the rapid industrialization of the South is changing the external aspects of things, the fundamental concepts of life and the techniques for realizing them remain basically unaltered. As conservative a figure as Mr. George Morris of the *Memphis Commercial Appeal* writes, "The only trouble with the South is that it is bound down by tradition. In a world of change it is satisfied with things as they are, notwithstanding they are pretty much the same as they were a hundred years ago." [2] This is due primarily to the tenacity with which an agrarian society clings to its way of life long after those specific patterns have lost their validity. It is the plantation, as Professor Charles Johnson has so admirably pointed out, that has cast the mould for southern civilization. It is not so much the plantation *per se* today that plays the significant role as it is the psychology of the people, their habits, manners, traditions, etc., which the plantation created and which now shadows the whole of southern life. The pattern of conduct which the plantation early laid down for the governance of poor whites and Negroes is today reflected not only in the attitude of present-day plantation owners toward their "hands," but in the mills, mines and factories which are becoming increasingly familiar in the southern scene.

The sharecropping system as we know it today has its roots planted deep in the historic soil of the South. While it was born of the cataclysm which befell the Southern Confederacy at the close of the Civil War, its pattern was cut and fashioned in the Colonial Period when America was young and vigorous. The plantation existed before sharecropping and even before chattel slavery. It was the colonial plantation which called for hordes of cheap laborers to feed, clothe, and delight the world. The early colonial plantations were worked by indentured white slaves who might eventually, by dint of hard work and the benevolence of the overlord, gain their freedom. The culture of cotton, rice, and sugar was strenuous and difficult work. Thousands of poor men died in the rice swamps and on the cotton and tobacco plantations. In 1619 a handful of Negroes were landed at Jamestown. From that time ships bearing slaves fed their human cargoes to the ever-increasing acres of King Cotton. On the sturdy backs of the strong black men from Africa, King Cotton reared his dominion to supremacy. Slave labor eventually became the imperative of southern economy. The poor whites were driven from the broad and fertile bottom lands and delta regions into the barren hills where they developed a deep hatred toward the planting aristocracy and the enslaved Negro who, through no fault of his own, had been ruthlessly uprooted from his native land, transplanted to the New World and forced to take the jobs which the poor whites wanted and needed. Andrew Jackson gave these disinherited white men some notion of their rights and a longing for more power with which to crush their overlords. It was not until the advent of men like Tillman, Blease, Vardaman, Bilbo and Huey Long that this rejected class tasted power and wielded it. But the psychology of the plantation had done its work. It had created in them the same fundamental responses for which the aristocracy had been so cordially hated. With the opening of the South to a predatory capitalism, young and ruthless, the masses of these disinherited whites along with the millions of "emancipated" Negroes became the unknowing victims of a parasitic, brutal and ugly society.

The fundamental question at issue between the North and the South was whether "slave" labor or "free" labor should prevail in America. Theoretically "free" labor won in the struggle, but basically the system remained unchanged and there are today millions of totally disinherited men and women, white and black, who are anxiously waiting for the day when a genuine emancipation will free them from a tyranny often more cruel and debasing than chattel slavery.

Lincoln's *Emancipation Proclamation* and Appomattox demolished the form but not the nature of southern society. The house was toppled over but the foundations remained for the most part intact. On top of the old, as was natural and human, the South began to build. Her people remembered much, forgot little and continued to live pretty much as they had before. The old ways of thinking and acting asserted their power with force and vigor. The plantation aristocracy still had their vast acres as well as their ideas as to how society should be organized and run. The four and one-half million Negroes had a kind of "freedom" and that was about all. They had been "emancipated" and left there. There were no "forty acres and a mule." Had there been, the subsequent history of the South would have been fundamentally different.

The bitter feeling which has existed between the poor whites and the Negroes is well known. It was a bitterness born of economic frustration of the poor white farmers who were crowded out of the rich bottom lands by the great planters. While they bore a deep resentment toward the planting aristocracy, the Negro became the object of their bitterness and scorn. The planting aristocracy, sensing the potential strength of the poor whites and the rising middle class, directed the grievances of these classes from themselves to the Negro masses who, in time, became the scapegoat for all of the South's sins. The cry of the poor whites was to "git the niggers off our necks." It was perhaps truer that they were a-straddle each other's necks and did not know it. The well-to-do planters and ruling class generally kept the poor whites too busy fighting the Negro to ask the one basic question, which sooner or later had to be asked. That question was: "Who put the Negroes on our necks?" The obvious answer was, "The planters, landlords, merchant princes, business men and all the others who derived a benefit from the struggle the black and white workers were carrying on between themselves." As long as these beneficiaries could keep the workers divided they could play one group against the other and in that manner defeat both. It was the old policy of divide and rule. For years this struggle made the exploitation of the workers in the cotton fields a simple matter. On a plantation in Poinsett County, Arkansas, a handful of white and black men were to ask the question and together get the right answer.

Inter-racialists of the Atlanta School take particular pains to point out the ancient hatred which has existed between the poor white man and the Negro. At the same time they take great delight in attempting to show that the rich man with his vast benevolence and paternalism is the Negro's best friend, conveniently forgetting that if the poor white man is the Negro's worst enemy it is the members of the so-called "best families" who force these equally exploited groups to struggle against each other. We have been told that the hatred the poor white man bore the Negro was so vile and intense that nothing, howsoever fraught with significance for either or both of them, could ever join their hands in a united struggle. The Southern Tenant Farmers' Union is a decisive answer to history and to the segregationalists. Black men and white men *have* joined together in a common cause against a common oppressor. They have done so of their own free will. No "outside agitator" told them what to do. Experience and hard common sense was their main guide. To the everlasting credit of the common man—from official Washington's point of view, the commonest of all men, the southern sharecropper—let it be said, that when these black and white men joined hands in fellowship and struggle, they exemplified a degree of wisdom and moral courage rarely witnessed in America.

Appomattox left the South prostrate and demoralized. The mighty landlords and the land remained, and Washington notwithstanding, Cotton was still King. The landlord-planter had his vast acres of land but no "hands," the "emancipated" Negroes had hands but no lands. Those who had and those who had not struck a bargain. If the landlord planter would furnish the land, a house to live in, food, tools and mules for the worker, the worker would in turn cultivate his crop and at the end of the season they would share the crop between them. At that time it looked like a just and workable scheme that would benefit everyone. A people floundering in a morass seldom indulge in futuristic thinking. They seize upon whatever is offered them as a way out. The Negroes and eventually the poor whites began sharecropping as a means of deliverance. The landlord's acres blossomed in white and there was a note of gaiety in the southern air once more. King Cotton was established on his throne again. Sharecropping had been a deliverance for the heirs of King Cotton, but for the sharecropper what seemed just and workable during the chaotic and black seventies and eighties had re-enslaved him once more. For nearly three-quarters of a century sharecropping has blighted and poisoned the whole of southern life.

By 1880 sharecropping as a system of working agricultural products was fairly well established. On the large plantations and small farms, this semi-feudal mode of producing cotton, tobacco, rice, sugar and other products was accepted as the best possible system for the South. Despite the fact that those who produced cotton were among the most wretched of all workers in America the system continued to expand. Each year thousands of men were added to the sharecroppers. They were in a class by themselves, disinherited, enslaved and unloved by those who wove their destiny.

By 1900, 45% of all farms in the cotton states were operated by tenants; by 1910 the number had grown to 50%; by 1920 it had mounted to 55% and today it is estimated that between 60 and 70% are farmed by tenants. Dr. Rupert Vance places the number of tenant families in the cotton belt at 1,790,783. Of these 1,091,944 are white and 698,839 are Negro. The number of individuals included in this classification runs to approximately 5,500,000 whites and 3,200,000 Negroes.

"My God, how do these people live!"[3] exclaimed a southern writer in 1880.

Thirty years later a prominent cotton speculator wrote: "American cotton planters (workers*), proprietors of the greatest gold-producing staple in the world are poor. They are in practical servitude . . . and the poorest paid toilers in the United States. Our greatest asset is our greatest humiliation. Cotton is king but it is a badly served monarch. . . . It does not enrich but rather impoverishes the Southland. An enormous profit is made somewhere in the progress of the cotton to consumer. Every year cotton goods of the value of nearly six billion dollars are turned out from the 125,000,000 spindles in the world. But the poor farmer in the cotton fields sees but a pitiful part of the multiplying fortunes attending the migration of cotton goods around the earth. . . . The working farmer for his product gets, we will say, ten cents a pound or fifty dollars a bale; his twelve months of effort and expense bringing him a gross revenue of $250 (twenty acres producing ten bales*). This is an insignificant total for the men who among others produce the commodity that controls the world. Out of that $250 he must provide for his family, himself, and his mule, and make provision for the ensuing times of planting and cultivating.

* Author's explanation.

Fully sixty-five per cent of American cotton crop is produced by this struggling method." [4]

A prominent South Carolina planter said in 1909, "The South hasn't had enough pay since the Civil War to advance a single step in civilization. Whole families lived on twenty-five cents a day in the years of five-cent cotton. . . . Perhaps we haven't all realized how wretched an existence the small cotton grower has been forced to lead. Since the Civil War, whether white or colored, he has been the equivalent of slave labor." [5]

"Whenever I travel through the Cotton Belt east of the Mississippi River the things that haunt me most are the ragged tenants, wrinkled wives, half-fed, anaemic children and the wretched hovels in which they live, whether white or Negro. For the past fifty years cotton has been produced out of the very life-blood of the South." [6]

In 1928, during the heyday of "Coolidge Prosperity," an indignant voice was raised against the terrible conditions under which the tenants, sharecroppers and day laborers in the cotton country lived. Writing in the *Dallas News*, T. N. Jones, of Tyler, Texas, said: ". . . In the South there is more abject poverty and illiteracy than in any other country on earth in which a high state of civilization is supposed to exist. The squalid condition of the cotton raisers of the South is a disgrace to the southern people. They stay in shacks, thousands of which are unfit to house animals, much less human beings. Their children are born under such conditions of medical treatment, food and clothing as would make an Eskimo rejoice that he did not live in a cotton growing country. Without exception around these shacks there are no decent sanitary accommodations. There are no places for the production and care of live stock or poultry. In hundreds of instances there are no arrangements for gardens or places where vegetables can be raised. . . ." [7]

Writing in 1935 in *Today*, Fred Kelly says: ". . . It is in Arkansas that one finds the situation of the sharecropper really tragic. I have never seen living conditions on lower standards, even in backward sections of Europe." [8]

Rev. Alfred Loring-Clark, a well known Episcopal clergyman of Memphis, Tennessee, in a letter to President Roosevelt, has the following to say:

"Year after year the South produces cotton. Hundreds of millions of dollars have been made on the South's cotton plantations. Fortunes have been made and lost by planters. Yet generation after generation of tenant farmers come and go while their economic and social conditions remain unimproved.

"The sharecropper of today is no more literate, no more wealthy, no more cultured, no more privileged than was his grandfather of fifty years ago. Obviously there have been enormous profits in cotton, yet the tenant farmers' share has been so small that he must struggle desperately to keep body and soul together. Hope, ambition and incentive have through the years been killed in those who till the soil.

"It is stupid to condemn them as idle, shiftless and immoral. If they are so, the responsibility is not theirs. The tenant farmer is today the product of a vicious, stifling environment into which he was born. He is condemned at birth to grow up in a poverty-stricken home, guided by illiterate parents who themselves are without ambition, without incentive. Idealism is a word unknown. Such a child matures, marries and in turn establishes a home no better, no worse than his father's. So generation after generation, the tragic fate of the tenant farmer is repeated. Only now does the South begin to realise that it must solve the problem of the tenant farmer." [9]

It is not necessary to register further indictments against King Cotton. The above citations, which could be extended *ad infinitum*, should be sufficient to convince even the most ardent Tory that something is drastically wrong in the cotton country. Why then is it that cotton has continued to be produced at such frightful human costs?

"The mill owner and the weaver," says a prominent cotton speculator, "want cheap cotton. Tenant farmers are under an enormous handicap and the landowner knows that high prices mean ability on the part of the tenant to get out from under the yoke and become independent; this the landowner does not wish." [10]

Mr. John A. Todd, well known authority on the world's cotton crops, says that "cotton has always been regarded as a cheap-laborer crop, and cannot be grown at a profit if all the labor . . . has to be paid for." [11]

Thus it is seen that the demand for cheap cotton by mill owners and manufacturers, and the desire of the landlord to prevent the tenant and sharecropper from becoming independent, and enlarging the field of competition, and consequently cutting in on his profits, works to defeat the economic rise of the cotton workers.

Those who profit most from the trade in cotton tell us plainly that in order for them to have their profits cotton must be produced in part by slave labor, that is, labor which receives no wages.

In order that excessive profits may be taken from the trade in cotton the entire cotton oligarchy combine to force into the labor market an abundant supply of labor, "cheap and docile." By having an abundant labor market from which to draw cotton field workers the planters are ever in a position to dictate the terms upon which the crop is to be raised. Consequently, the ever-growing number of cotton workers makes them increasingly at the mercy of the planter, and places in his hand a powerful club. When any single tenant or sharecropper, or any large group, attempts to bargain with the landlord for better terms, he is continuously conscious of a vast army of unemployed men hungry for their places. The planter, well aware of this condition, either drives his workers to accept his terms or evicts them without law or ceremony from his place.

It is beyond doubt that the traditional attitude so assiduously cultivated by the "best families" that "anything is good enough for poor whites and niggers" has worked increasingly to lower their already miserable standards of living and greatly to undermine their morale. Such attitudes have helped to drive into a state of abject poverty and despair an enormous segment of our population, white and Negro.

Another item that must be included in the reasons why the sharecroppers and tenant farmers have always been in such desperate circumstances is the single crop system of the South. Cotton is a cash crop and planters want cash for what their tenants produce. The modern agriculturists can talk about diversification until they are blue in the face but as long as the planters are in complete power to do as they choose about crops no real change will come about. The tenant may want to diversify and have a garden but when the landlord tells him to plant cotton and furnishes him with no other seed he is powerless to do anything but plant cotton.

Finally, it may be said that there can be no great change in the status of the millions of sharecroppers, tenant farmers and day laborers in the South as long as the present economic system endures. Only a society that plans for production for use and not for profit and puts an inestimable worth on human beings can greatly alter the tyranny and terror in King Cotton's South.

The traveler in the cotton country will realize, if he gives an eye to the toilers in the fields, that cotton is produced "by the very life-blood of the South." He will not see many of the antebellum plantations for most of them have been broken up into small tracts, but a few remain and where some have disappeared countless others have arisen to take their place. Thus it is that great plantations exist today where once independent small farmers lived and cultivated their own land. But through illegal extortion, legal manipulations, mortgages and foreclosures, literally thousands and tens of thousands of poor farmers who once tilled their own acres have passed successively from the status of owner to tenant farmer and from tenant farmer to sharecropper and from sharecropper to the casual day laborer. Most of them have been "squeezed out" by the large plantations, insurance companies and banks, and each year they swell the ranks of the dispossessed sharecroppers combing the South for a "just landlord and some kind of a living."

Today the traveler will see them on the highways on "the move" always in search for "something better" which they rarely ever find. Pulling their earthly possessions on a makeshift cart, the disinherited and his family search until they find some place to make a crop and sleep at night. Often there is no place to make a crop but only an abandoned barn or shack to afford them scanty shelter. One sees them on the rivers in flat boats, in the coves, swamps and on the barren hillsides—a great mass of wanderers, without homes, food or work, half clothed, sick of body and soul, unwanted and unloved in the Kingdom of King Cotton which they have faithfully served but which no longer needs them. If the traveler pauses long enough to have a "look in" upon some cropper's home he will probably find two or three families "doubled up." Living in quarters barely sufficient for a family of three or four, a sharecropper who has a house to live in and a crop to work flings wide the doors to his less fortunate brothers who have neither home, crop nor job. Thus one may find three families crowded into a shack which could ill afford to give comfort to one.

And who are these people, these wanderers without friends or shelter? And who are these their brethren, who, possessing next to nothing, give them the only security and fellowship they know? They are the multitudes the planters drove from their plantations when Secretary of Agriculture Henry A. Wallace perpetrated upon the South the greatest iniquity since the "bo' evil," that economic monstrosity and bastard child of a decadent capitalism and a youthful Fascism, the AAA.

The "plow-up" program of 1933 and the reduction program of 1934 not only reduced the amount of cotton that was being produced, but "plowed under" hundreds of thousands of sharecroppers in the South and reduced them to an all-time record of poverty, misery and hopelessness. Not only were cotton tenants eliminated from the land by the reduction program, but those who remained on the land had their already miserable standard of living lowered and received practically none of the benefits which were due them under the government contracts. Millions of taxpayers' dollars have been poured into the hands of planters and absentee landlords and there most of it has remained. It is only fair to say that some landlords have kept the contracts with their tenants but this is the exception rather than the rule. Intolerable were the burdens the sharecroppers were carrying upon their frail shoulders before the advent of this messiah of economic scarcity who merely succeeded in piling higher the burdens they had had bequeathed to them out of a long and gloomy past.

With a picture of the normal conditions which confront tenant farmers, sharecroppers and day laborers in the cotton fields of the South in our mind, let us proceed to view the present scene as painted by the AAA.

However well meaning the present administration of the affairs of the Department of Agriculture may appear to the casual observer, it is a fact recognized by practically every competent economist that the actual working-out of the AAA in the cotton country has had a calamitous and devastating effect on the masses of people, white and colored. Indeed, it is doubtful if the Civil War actually produced more human suffering and pauperized more individuals in proportion to the population than the AAA has done in its few years of existence.

Professor Harold Hoffsommer of Alabama Polytechnic Institute, says, "The whole action of the AAA to the present time appears to have the effect of not only maintaining the status quo in landlord tenant relations, but of actually strengthening the foundations upon which they are built." [12]

One could take seriously the many public utterances of the Secretary of Agriculture, which have at times the earmarks of a genuine sense of social justice, were it not for the fact that he has turned a deaf ear to hundreds of genuine grievances laid before him; has permitted his chief assistants to whitewash appalling conditions and violations of government contracts they themselves knew about and investigated;

permitted most of the socially-minded men in his department to be ruthlessly kicked out of office because they had serious intentions of administering the contracts with justice and fairness to the sharecroppers and tenant farmers; personally liquidated the now famous but still unpublished "Mary Conner Myers Report" on conditions in the Arkansas cotton country; has chosen to ignore the fundamental issues in what he himself admits is a "sore spot" and blames "socialistic and communistic gentlemen" for the seriousness of the situation in the Mississippi Valley, and has referred to southern-born and educated men who have been working among the sharecroppers as "religious sentimentalists who know nothing about the South."

In keeping with the post-war trends of nearly all western European governments to shape their national life around an economy of scarcity, the Department of Agriculture built the agrarian policies of the Roosevelt "New Deal" Administration in the cotton country around a drastic reduction of cotton. I am thoroughly familiar with the arguments advanced by the eminent gentlemen in Washington as to why such a policy is, at this particular period of our national life, necessary and justifiable, but no amount of finely spun argument will feed the naked, clothe the hungry and give jobs to those who have been driven off the land because of the reduction program. It is true that the export market for cotton has had a steady and at times precipitous decline; it is likewise true that the domestic consumption of cotton goods has never been sufficient to allow a minimum of decency and comfort to exist on the bodies and in the homes of the masses of toiling workers in America. The average per capita consumption of cotton goods, including clothes, sheets, towels, pillow cases, etc., is only nine pounds. Instead of reducing the amount of cotton raised, an intelligent and socially-minded administration would have increased the production at least eighty or ninety per cent. But Bernard Shaw has well observed that the earth is the universe's lunatic asylum: had he known America better he might have added that the chief lunatics dwell in Washington.

In 1933 the South was treated to the "plow-up" program when the producers agreed to plow up between twenty-five and fifty per cent of their acreage for which they were to receive a rental payment of approximately $11 per acre. It is estimated that 4,400,000 bales of cotton were plowed under in 1933.

A lot has been said about the difficulty cotton farmers had in getting their mules to step on the cotton. The mules had better sense but like sharecroppers they are employed for their brawn and not their brains. The "plow-up" program was followed in 1934 by the reduction program when forty per cent of the land normally under cultivation was retired. This land was rented by the government and payment was made direct to the landlord or planter.

Section 10—the decision of the Supreme Court of Arkansas notwithstanding—of the Cotton Acreage Reduction Contracts guarantees to the tenants and sharecroppers the right to share in the benefit payments. The ordinary sharecropper without tools or teams of his own was allowed one-half cent per pound for cotton not grown in 1934 as his share of the "parity payment" or as the sharecroppers themselves call it the "poverty payment." The owner received all of the rent money, three and one-half cents per pound, and one-half cent of the parity payment. This, as Dr. William Amberson observes, is a very "curious eight to one division" in favor of the planter. Dr. Paul Bruton, formerly of the legal staff of the AAA, has this to say about the division: "The contract should have been drawn so that the benefit payments would have been made directly to the landlords and tenants in proportion to their respective interests in the crop. . . . *Under the 1934 and 1935 contract the landlord has everything to gain and the cropper everything to lose.*" [13] (Italics mine.) As the situation worked out the landlord did gain everything and the sharecropper lost practically everything.

In Section 7 of the contract we find the further rights of the tenants set forth.

> "The producer shall endeavor in good faith to bring about the reduction of acreage contemplated in this contract in such a manner as to cause the least possible amount of labor, economic and social disturbance, and to this end, insofar as possible, he shall effect the acreage reduction as nearly ratable as practicable among the tenants on this farm; shall, insofar as possible, maintain on this farm the normal number of tenants and other employees; shall permit all tenants to continue in the occupancy of their houses on this farm, rent free for the years 1934 and 1935 (unless any such tenant shall so conduct himself as to become a nuisance or a menace to the welfare of the producer); during such years shall afford such tenants or employees, without cost, access for fuel to such woods land belonging to this farm as he may designate; shall permit such tenants the use of an adequate portion of the rented acres to grow food and feed crops for home consumption and for

pasturage of domestically used live stock; and for such of the rented acres shall permit the reasonable use of work animals and equipment in exchange for labor." [14]

It is clear that some thought was given to the protection of tenants and sharecroppers from displacement from the land. However, a close examination—a critical examination is hardly necessary—reveals the essential weakness of the document written in favor of the propertied interests throughout. The producer will "*endeavor* . . . to bring about reduction as to cause *the least possible* amount of labor, economic and social disturbance, and . . . *insofar as possible,* he shall effect the acreage reduction *as nearly ratable as practicable* among tenants, . . . shall *insofar as possible,* maintain . . . *the normal* number of tenants, shall permit all tenants to continue in the occupancy of their houses . . . free . . . (*unless any such tenant shall so conduct himself as to become a nuisance or menace to the welfare of the producer*)." It is readily seen that the contract placed the "producer" in complete control of the enforcement of all or any part of the contract. He was under obligation to no one save county agents who were in most cases incapable or unwilling to see that the contracts were strictly observed. He (the producer) himself, in most instances, was the interpreter of whether he had reasonably fulfilled the contract. Only in instances of the most flagrant violations were they ever cancelled.

Dr. William R. Amberson, with a group of students, made one of the first, most thorough and valuable studies of the effects of the reduction program on the lives of tenants and sharecroppers that have been made.[15] The survey covered about 500 families in Mississippi, Tennessee and Arkansas. The Amberson Committee reached the following conclusions:

> "The acreage-reduction program has operated to reduce the number of families in employment on cotton farms . . . due . . . to the failure . . . to reduce acreage ratably, forcing some tenants into 'no crop class' . . . at least 15% . . . of all . . . families. . . . Many plantation owners eliminated the sharecropping system . . . forcing . . . croppers to accept day labor instead. . . . Widespread replacement of white by colored labor. . . ." [16]

What Dr. Amberson's committee discovered in their investigation has been subsequently established generally throughout the cotton country.

Despite the plain wording of Section 7 that the landlord shall ke
the "normal number of tenants and other employees employed on this farm" there have been wholesale evictions of tenants. After studying 700 displaced farm families in North Carolina, Gordon W. Blackwell comes to the conclusion that "the fact that the landlord could no longer finance the tenant, the desire of the landlord to use the tenant as day labor rather than give him a crop, and the acreage reduction program of AAA are the real reasons why there is a displaced-tenant problem." [17] Raymond Daniel, correspondent of *The New York Times*, found that scores of displaced tenants were on the relief rolls in Memphis and that a large number of them came from Arkansas. "We believe," writes Dr. William Amberson in *The Nation* for February 13, 1935, "it is fair to say that over the whole cotton belt about one-third of the present rural unemployment can be directly referred to the reduction program." [18]

A wave of protest arose over the mal-administration of the reduction contracts and the Department established an Adjustment Committee headed by J. Phil Campbell. Several thousand complaints were filed with the Committee and some adjustments made. The way in which the contracts were violated and the nature of the government's investigation of the charges is set forth in an article in *The Nation* by Dr. William Amberson. "Our committee," says Dr. Amberson, "observed the handling of cases submitted to Mr. Wallace in May. The committee found in one case that a large plantation had replaced many white sharecroppers with colored day laborers. The investigator reported that 'there have been some evictions of white families, and . . . some substitutions of sharecroppers with day labor, but the extent . . . has been very small in proportion to the size of the operations'; as a result the plantation was cleared. In another case 'it was found that a change had been made from sharecroppers to day laborers, but as the croppers had been notified early in 1933 that this arrangement would be carried out in 1934, it was not considered that this change was made as a result of the cotton program.' The committee (Amberson Committee) holds affidavits from some of these people, all white, in which they swear that they received no notice until after January 1, 1934. Similar affidavits were submitted by them to the government investigator but were ignored." Continuing, Dr. Amberson says: "A large cotton farm in eastern Arkansas, comprising some 14,000 acres, has long been notorious for its bad treatment of its tenants. . . . Interest on 'furnish' has been charged at twenty-five cents on the dollar—an illegal and usurious rate—when settlements were made, and few settlements have been carried out. The condition of the croppers, mostly colored, has been tragic. In one day last winter the FERA worker in the county, aghast at the condition of these people, spent $1,400 of government money to clothe and feed them. The only clothes which most of them now possess were given them by this worker. In the spring of 1934 the owners decided to change a part of this farm over to a day-labor basis. . . . In some cases this represented a cut of seventy-five per cent in acreage. They were required to cultivate a large part of the plantation on a day-labor basis at seventy-five cents for a thirteen-hour day. Actually thirty-five cents only was paid over; the rest was placed in a 'petty account' which the croppers claim has not been paid. The government investigator went to this plantation with a member of the County Committee (government), who has informed our committee that no thorough investigation was made. A few croppers were interviewed and a few questions which were not pertinent to the charges were asked. Thus the investigator chiefly inquired whether the families had enough corn land. These people were rightly suspicious of all inquirers, and they failed to disclose the real situation. The investigator conferred with the owners but made no inspection of the books. *He told his guide that he thought he had not been told the truth. But no word of this opinion appeared in his report to Washington, which cleared the plantation. The AAA check for thousands of dollars was shortly thereafter released, and the records state that 'this plantation was thoroughly investigated.' The whole 'investigation' of this huge farm covering twenty-two square miles was completed in not more than six hours.*" [19]

The cases cited above describe the situation better than statistics.

The cotton control program was written in behalf of the landlords and planters and is correctly known as "the landlord's code." In the Commodity Information Series the question is asked, "Where does the responsibility for the administration of the cotton program rest? Answer. The administration is primarily local, resting upon the community and county committees chosen by the cotton producers." [20] Since the control of the program was lodged in the hands of the landlords it is rather obvious that whatever the tenants and sharecroppers got out of the program depended upon their relationship to the landlord. The record shows that the relationship was a poor one for the majority of tenants and sharecroppers.

One of the obvious flaws in the program was the failure of the contracts to provide for the representation of tenants and sharecroppers on the boards of control. The inevitable result of the failure so to provide was that the rights of tenants and sharecroppers were completely ignored or evaded in one way or another.

The government has not only given the landlords and planters their own code but has strengthened their hands in other ways. In a depressed world market the AAA makes the planter a fancy price on his cotton. At the same time it pays him for not planting a certain percentage of his land. Tenants under the contracts were legally entitled to use this government rented land rent-free but many a landlord with both eyes to his profits made tenants pay for the use of the land and evicted them when they could not. While the landlord received the lion's share of the rental money, the tenants suffered an incredible hardship in that their acreage was greatly reduced and their standard of living greatly lowered.

Through the Federal Credit Administration the landowner can get loans at 4.5 per cent to 6.5 per cent, whereas the tenant has no access to such low rates of interest and the landlord who is being generously assisted by the government goes right on charging the tenant the customary ten cents or more on the dollar.

Rarely, if ever, did the tenants receive cash payments for their part in the reduction program. The checks went to the landlord and were credited by him to the tenants' accounts at the commissary. If there was not a commissary handy the county agent frequently distributed the checks at the credit stores. Here the checks usually went to the merchant for past, present or future "furnishings."

Without further burdening the reader it is sufficient to say that the program of the government was a great success from the planters' viewpoint and an utter and miserable failure from the viewpoint of tenants and sharecroppers.

In company with a small group of sharecroppers from Alabama, Mississippi and Arkansas, the author went to see Mr. Cully Cobb in Washington in regard to their parity payments which none of them had received. Mr. Cobb tried to coerce one of the more intelligent of the group to state that some of the people in her district had received their parity payments. "No sir, ain't none of us had no 'parity money.'" What Mrs. —— from Mississippi could say about the nonreceipt of parity payments nine out of ten sharecroppers could say.

For every tenant who has received his share of the government money without endless delay one hundred have never received anything.

Tenants who owned their own mules and tools and worked for a landlord were supposed under the contracts to receive part of the rent money the government was paying the landlord. Some of the charges recently made are borne out in a typical case cited by Raymond Daniel of *The New York Times*. "Chapman-Dewey moved some twenty-five families . . . among the cases was that of C. G. Fletcher, a sharecropper on the Green and Reynolds plantation in Craighead County. Having his own team and tools, Mr. Fletcher was able to rent cotton land . . . for one-fourth of his crop, but he was charged $7 an acre for land on which to grow corn and hay for his stock. Since his landlord was receiving rent from the government in connection with the acreage reduction program, Mr. Fletcher thought he was entitled to share in the government benefits and sought to have himself declared a 'managing share-tenant.' Mr. W. J. Greenfield, representative of the Department of Agriculture, investigated Mr. Fletcher's case, similar to that of many others, and held that, since his farming was supervised by a riding boss, he was not a managing share-tenant within the meaning of the law and therefore was not entitled to any part of the government rental. On December 21st, 1934, Fletcher was ordered to vacate his house by January 1, 1935." [21]

The Bankhead law ordered that all bale tags should be in the possession of the person who actually produced the cotton. This was a departure from the old way of doing things for always before the tenant had turned his cotton immediately over to the planter who had it ginned, baled, and, when he received the money, applied it on the tenant's account. Landlords in Arkansas early began to force tenants to sign papers making them the trustees of the bale tags. Some tenants refused and were evicted.

While sharecropping is the devil's brew, it is a life greatly to be envied by those who must win their bread by day-labor in the fields. Sharecroppers, as Norman Thomas has observed, are men who possess as nearly nothing as any people in the United States, but day-laborers possess less than what the sharecroppers do. They work when they can, at what they can, wherever they can. They accept—are forced to accept by economic necessity—whatever the landlord desires to give them. It is cheaper to work a plantation by day-labor than by sharecropping and today the trend in the cotton country is swiftly moving in that direction.

The relief machinery of the federal government is all too frequently used as a club to force the unfortunate day-laborer and sharecropper into accepting arduous and difficult labor in the cotton fields at less than starvation wages. In a county in northeastern Arkansas in which the Southern Tenant Farmers' Union was particularly active and where the planters were changing from sharecropping to day-labor, the manager of one of the largest plantations in that area made daily trips to the county relief headquarters. Each day he brought back with him the list of persons who were on the relief rolls. To these he and his associates offered work in the radish fields at twelve and one-half cents to twenty-five cents per day. They worked ten and twelve hours. The command was to accept this work at this wage or get off the relief rolls. Whether the man worked or did not work it made relatively little difference. His stomach was just as empty and his body just as naked whether he worked or not. Once again the federal government sided against the disinherited in making them slave in the fields for those who were in a position to exploit them.

Although tenants on many plantations were suffering acutely because of the reduction program and were in desperate need of food and clothing, hundreds of landlords were opposed to their tenants receiving government relief for fear it would make them less dependent upon them and create in the tenants a desire for better living conditions.

Writing of these same people in July, Frazier Hunt, correspondent for the *New York Wold-Telegram* says: "Bad food, unsanitary homes, climate, ignorance, disease and hopelessness had told on their faces and bodies. They seemed to belong to another land than the America I knew and loved." [22]

Disinherited, oppressed and enslaved, these men who "seemed to belong to another land" are murmuring against their overlords. Today there is a rumbling in the cotton country: King Cotton's slaves are stirring.

III. THE SHARECROPPER

"He has no such rights as peasants had acquired, even in the Middle Ages under Feudalism."

—Norman Thomas

"The landlord has the upper hand. He owns the stock, the implements, the land, and sometimes by an unwritten law, the cropper himself."

—Oliver Carlson

"My God How do these People live?"

—North Carolina Attorney

In our world cotton has become a first necessity. Let the reader pause for a moment and try to picture for himself a world without cotton.

In the production of cotton—as is true of coal, iron and most of the things to which we moderns have become accustomed—rarely, if ever, are we conscious of the *human factors* involved. Of all the agricultural workers in America none are more exploited, brutalized and degraded than the cottonfield workers of the South. For all our talk about democracy and the high standards of living of American workmen, virtually millions of our fellow-citizens are living in practical slavery. To call these workers in the cotton fields "slaves" is not to indulge in wild or fancied statements. The learned professors and the sophisticated college students will choose to describe their status as that of "peonage." The men and the women who toil in the fields care nothing for our fine reasoning and the traveler among them today will find them referring to themselves as "slaves" pure and simple.

The sort of slavery to which they refer has been dignified by the Chamber of Commerce as free labor, that is, labor that is free to take what the landlord-planters want to give when the crop is harvested.

It is a kind of slavery that is, in certain aspects, more vicious and devastating than chattel slavery and few things are more abominable in the sight of God or man than chattel slavery. Today you will find hundreds of thousands of cotton workers who have none of the benefits that were not infrequently generously extended chattel slaves in the Old South. Under slavery a man was of some value. He was usually provided with a fairly good house in which to live, sufficient clothes to keep him warm and comfortable, and enough food to nourish his body. When he became sick a physician attended him. What about the modern sharecroppers? Thousands live in shacks unfit for animals. Their diet consists mainly of cornbread and molasses. Rarely does one find a cotton farmer who has sufficient clothes to keep his body warm and comfortable. If he becomes ill and has money he can pay a doctor; if not, he can die. The South is a land of "cheap labor, plentiful and docile" where a sharecropper is worth less in the eyes of many a landlord-planter than a good mule.

The sharecropper "is a man," says Norman Thomas, "who owns, on the average, as near to nothing as any man in the United States. . . . He has no such rights as peasants had acquired, even in the Middle Ages under Feudalism." [1]

What then is the real status of a sharecropper in modern America? The authors of *The Collapse of Cotton Tenancy* have given a remarkably succinct and pointed description of the status of the sharecropper:

"The status of tenancy demands complete dependence; it requires no education and demands no initiative, since the landlord assumes the prerogative of direction in the choice of crop, the method by which it shall be cultivated, and how and when and where it shall be sold. He keeps the records and determines the earnings. Through the commissary or credit merchant, even the choice of diet is determined. The landlord can determine the kind and amount of schooling for the children, the extent to which they may share benefits intended for all the people. He may even determine the relief they receive in the extremity of their distress. He controls the courts, the agencies of law enforcement and, as in the case of sharecroppers in eastern Arkansas, can effectively thwart any efforts at organization to protect their meager rights." [2]

The object of this study is to portray the life of the Arkansas sharecropper and the struggles of the Southern Tenant Farmers' Union against the domination of the plantation overlords. We make no attempt to draw any sharp distinction between tenant farmers and sharecroppers for in our opinion the difference between them is negligible, and of little significance, if any at all. What the writer has to say about the Arkansas sharecropper may be said about sharecroppers in general throughout the length and breadth of the South. The "system" operates about the same whether it be in the Carolinas or Mississippi. In order that the reader may not think that the description in the following pages is wild and extravagant language, I shall quote from a conservative source. Mr. J. T. Holleman, a Director of the American National Bank of Atlanta and President of the Southern Mortgage Company, wrote in the *Atlanta Constitution* of September 27th, 1914: "Pathetic indeed has been the life of the small landowner and tenant farmer in Georgia and the South for fifty years. Courageous, honest, patient, and long-suffering. . . . In the springtime they go forth with their brothers in black, set their hands to the plow. They bend their backs to the burdens and when the frost falls they have added $1,000,000,000 to the wealth of the world. But small indeed is their share, and meagre is the recompense to them. Every two years . . . they move from one place to another. They build no homes, they live in rude huts, no flowers about their dwellings, no trees to shade them from the sun, consumed by the summer's heat and chilled by the winter's cold, no lawns about their houses, no garden fences, and with the accursed cotton plant crowding the very threshold of their rude dwellings." [3]

The sharecropper is an agricultural worker who has only his labor to sell. He may work on a large plantation of five, ten, twenty or thirty thousand acres, or he may work on a small farm. His landlord may hire him directly or he may hire himself to a planter who rents the land from a landlord who rents the land from a real estate dealer in Kansas City who leases it for a syndicate in New York and whose members may be scattered over half the world. Thus on the back of one man four or five others may be living in luxury while he himself toils and sweats usually for his keep, or as Norman Thomas has so aptly said, "he farms for exercise."

If the sharecropper works on a small farm at which the owner lives and works also, his chances of "making a livin'" are better than if he were located on a large plantation. It is generally true that the exploitation of the cottonfield worker is in direct proportion to the size of the plantation. The larger the plantation the more vicious the exploitation of the workers. The owner of a large plantation will probably find life in the city more to his liking, and he will become that abomination of modern life known as the absentee landlord. In his absence an overseer or manager is placed in charge of the plantation.

The manager's job is that of overseeing the work on the plantation and is usually dependent upon his ability to make the plantation pay high profits to the owner, and this frequently depends upon his willingness to exploit the tenants. Under the manager are riding bosses, whose duty it is to superintend or boss a gang of workers on a particular section of the plantation. They usually receive a fixed wage. These men, the riding bosses, are the terror of the plantation. They are the modern equivalent of slave drivers. They are usually uneducated and more often than otherwise are brutal, ruthless and domineering. Nearly all "tote guns" and are often the sheriff's deputies. They maintain with whatever vigor is necessary the iron code of the plantation. They may commit crimes for which the average citizen would be swiftly punished but they are "the law." As an Arkansas sharecropper once aptly observed, "Riding bosses, they ride horses and sharecroppers."

A visitor to the cotton country is likely to be awakened from a peaceful southern slumber by the sound of a distant bell ringing in the early hours of the morning, before the sun has begun its upward climb. The visitor will probably say to himself: "How lovely. This is a church bell calling the people to worship before they begin their day's work." The tones which the bell gives out may inspire the stranger in the cotton country toward reverential thinking for somehow a ringing bell reminds one of God and cathedrals and a man's eternal quest for truth and justice. But the bell which rings in the early morning hours on a cotton plantation brings no such thoughts to the mind of the tenant. Instead it brings forth thoughts of another day of unremitting and unrequited toil in the cotton fields.

The sharecropper will slowly drag himself out of bed. His wife has been up hours seeing about breakfast, and the children have left their pile of quilts on the floor, and are standing about the stove or grouped in the front door staring out upon the dreary, monotonous acres of cotton before them. On a rickety table they eat their breakfast consisting of biscuits, molasses, and fat-back or, as it is universally known in the South, "sow-belly." Hundreds of thousands of sharecroppers and tenant farmers have never known any other kind of breakfast and today they consider themselves fortunate to have even this. There are thousands, the sharecropper knows, who cry for bread and get none—who make daily trips to the county seat for some groceries and return empty handed, hungry, forlorn, hopeless. He has seen the children of these disinherited men eat clay. Today they eat clay, tomorrow snakes, the day after tomorrow roots of tender trees.

Another bell rings and the sharecropper and his family trudge toward the cotton patch. Here they work all day, pausing for a while at noon to partake of another meal of cornbread, molasses and meat, or perhaps, as is the case with so many today who cannot afford but two meals a day, just to rest from the terrible ordeal in the fields. When the sun casts its final shadows over the white acres, the family begins the weary walk to the place they call home. Tired, worn, and forlorn they huddle over the table scarcely lit by the kerosene lamp and eat cornbread, molasses and sow-belly. Soon the lamp is turned low. If there is a union meeting or some sort of religious gathering the family may find strength enough left to attend. But most likely the family will "pile" on beds or pallets on the floor, but not to rest, for men's bodies can find no rest on a corn-shuck mattress or on a heap of quilts and rags piled on a pine wood floor.

Of all the dreary sights in the cotton country the most pitiful are the shacks in which the tenants live. Of one, two and three rooms, they are probably the vilest places in which men have to eat and sleep in America. Writing in the *New York World-Telegram* for July 30th, 1935, Frazier Hunt says: "Scattered over the landscape were unpainted one and two room shacks. Many of them had no out-buildings of any kind. They were hovels, where human beings could eat and sleep. It would be a little unfair to the precious word to call them homes." [4] One can travel from Texas to Virginia and see these miserable abodes of cotton workers. Of clapboard construction, unpainted and weather-beaten, most of them are unfit for humans. If a sharecropper makes any improvements on the house in which he lives or on the property, he receives no compensation for his labor and the improvements go automatically to the landlord the moment the cropper leaves. In the sharecropper country of northeastern Arkansas one grows accustomed to seeing the work-animals housed in good barns and stables; but the human pigsties mar what little beauty there is in the landscape. But I have remarked already that sharecroppers in the cotton country are frequently looked upon by their overlords as of less value than a good mule.

When it rains the water pours through the walls and roofs of these hovels, necessitating the shifting about of the crude and ancient furniture. What furniture most sharecroppers have could be hauled in a one-horse wagon and, if the road were rough, it would fall to pieces. A few rickety chairs, a table, a bed or two, a few ragged quilts, what is left of a dresser or wash-stand, some broken boxes, a few broken dishes, a pig or a dog and once in a great while a few scrawny chickens usually constitute the sole possessions of a sharecropper.

The windows and doors of these human dwellings are rarely if ever screened, allowing a ready access to mosquitoes and flies from the unprotected out-house—when there is one. One can travel for miles in the cotton country of northeastern Arkansas and other heavy cotton raising states and never see an out-house. When a New England woman who had just purchased a plantation began to furnish her tenants with out-houses, she raised a flood of protests and disapproval from her planter neighbors. "You," they said, "are making our tenants dissatisfied. If you build out-houses, we will have to build them too. After all, Miss, all that a sharecropper needs is a cotton patch and a corn cob." This story may shock the sensitive but it illustrates better than almost anything I know the utter callousness of some landlords on the cotton plantations of the South toward the tenant farmer and sharecropper.

Malaria and pellagra are as common to many sharecroppers as their diet of meal, molasses and meat. From early childhood the sharecropper lives in dread of these twin plagues which continuously dog his daily existence, and when he dies his body is a mute testimony to the frightful ordeal through which he has passed. One reason for the prevalence of pellagra is that he lives, as Kroll has thoughtfully observed, in what is literally a "cabin in the cotton." Cotton crowds his very doorsteps and every inch of cultivatable soil must be planted in the precious fiber. There is no room, therefore, among cotton workers for a garden, where they might grow vegetables which would provide their families with healthful food for summer and winter. Cotton is King and he must be served irrespective of the human consequences. Under the present conditions of work, if a tenant had a garden, when would he find time to work it? A garden does not grow of its own accord, especially in a country where weeds flourish to choke out the life of the tender plants by which man must live. From sun-up to sunset, or as the cottonfield workers say in Arkansas "from can to can't," does not allow any time for working a garden. Furthermore, the planter does not want the tenant to grow vegetables or pigs or cows or chickens, for whatever the tenant raises himself he will not have to buy at the commissary. The planter wants his tenants to buy everything they need at the plantation commissary, where prices are high and profits easy. These three factors then—the demand of the planter for cotton, the lack of time during the growing season to cultivate a garden, and the desire of the planter to make profits at the commissary—account for the fact that as a rule tenants in the cotton country rarely have gardens.

In the company owned store and in the general credit stores the tenant is usually cheated and robbed. Here prices are often double what they are in independent stores. What he gets at the company store is "charged against him." He seldom asks the price of anything for it won't matter in the long run. He rarely keeps a record of what he buys and how much he pays for it, as it may bring on a dispute between himself and the manager of the commissary. Writing in the *American Mercury* for February, 1935, Oliver Carlson says, "He (the planter) it is who enters the accounts, and woe to the foolhardy cropper who dares question what has been entered on the planter's books." [5] The manager and the clerks charge what they like and always enough so that when the accounting is done at the end of the year there won't be anything coming to the tenant. A Georgia planter reveals the attitude of the average plantation owner better than I can state it: "*There is only one way,*" he writes, "*for a planter or farmer to make money with the use of hired labor, and that is to have a grab or commissary and keep books—always careful that no laborer exceeds his account, and making sure that at the end of the year he has gotten it all, and his labor has 'just lived,' as one would say.*" [6] (Italics mine)

It is true that out of the plantation commissary the expenses of running the plantation are often made. What the planter sells his cotton for is pure gain. At the commissary the workers are charged "sky-high prices" but they seldom protest, for protests may mean a beating or eviction from the plantation and after all a man "has to get by somehow."

In northeastern Arkansas the planters issue their tenants what is known as "doodlum books." The "doodlum books" or "do-you" books contain coupons worth from a penny to fifty cents. When a tenant wants something at the store the clerk takes the coupons in trade. The coupons are good for their face value only at the company store. The bank will not cash such coupons and to trade them at an independent store means that the tenant loses at least twenty-five per cent of their value.

Concerning the operation of plantation commissaries, Mr. Fred Kelly observed in *Today* for March 30th, 1935, "In buying at such a store or commissary, a tenant uses what is commonly known as a 'doodlum book'. . . . Prices for goods bought with these tickets are considerably higher than for cash. In one store, for example, I noticed that the cash price of potatoes was $1.75 but with a scrip book a buyer had to pay $2.25. Sometimes the difference runs as high as forty or fifty per cent. Moreover, the tenant using these advances must pay high interest—sometimes as much as twenty-five per cent. . . . I heard of several cases in which twenty-five per cent was added even though the advance had been made only for a day or two." [7]

Once the fertile land which now produces cotton reared on its belly giant forests. Today most of the wood has been cut and where once stood the mighty monarchs one can look for miles and not see a patch of trees. Lumber companies denuded the forests and then in order further to exploit the land, they turned to raising cotton on their vast acres. The Singer Sewing Machine Company operates a huge plantation on the site of a once great forest. The sharecropper, therefore, frequently has great difficulty securing firewood. It is a common sight to see men, women and children carrying wood on their backs from a distance of three and four miles. During recent years children have attended school, not to learn from books, but to get warm. There was a fire in the schoolhouse but nothing but a cold, empty stove at home. The tenants rarely see a lump of coal and often there is not enough wood to be had to keep them warm from the wintry blasts that blow through their warped and hole-ridden shacks.

Most Americans are accustomed to good drinking water. Not so the tenant in Arkansas. The water is frequently polluted with typhoid. The earth laps up the filth from the out-houses and barns. Good wells and cisterns are the exception in the cotton country and frequently there will be a "cabin in the cotton" with no water facilities at all. A single well, or a pipe driven into the ground may furnish a half-dozen or more families with their entire supply of water.

It is often said that the cobbler's children go without shoes, the butcher's children without meat and the tailor's children without clothes. It is true that the sharecropper, his wife and children who grow the cotton that clothes the world, are the poorest clad mortals in America.

In Arkansas, the saying goes, that the kind of clothes, shirts, towels, sheets, pillow cases, etc., that a sharecropper family has is determined by the brand of flour they happen to buy. It may be a "Gold Medal" pair of drawers the daughter wears or a "Red Cross" or "Grandmother's" shirt the sharecropper himself wears. The overall is the universal dress for the man. His wife and children do the best they can. I have seen scores of families with but a single piece of clothing for each member of the family. I have seen pregnant mothers without a single piece of clothing to give comfort to the newcomer. I have seen babies come into the world to be wrapped in old blankets, quilts and pieces of rags and I have seen them go out wrapped in no more. Near Harrisburg Corner I visited a former sharecropper and his wife who had just been evicted from their place on a plantation because of union activity. They were living on a raft upon which some rough lumber had been gotten together for a house. They were a young couple and their first baby was already overdue. There was a cot with a few torn quilts, an oil can with one end knocked out for a stove, two boxes for seats, some boards laid across a barrel for a table and some broken bits of dishes. The mother wore her *sole* dress. "The baby's coming and we haven't anything for it," the mother told me. When the baby came, there was no doctor, no midwife, no one but the father and mother to welcome it and try to give it comfort. In Harrisburg not ten miles away the Federal Emergency Relief had its headquarters. Six miles away was Marked Tree with its planters, preachers and general towns-people. When some used clothing was given to this family, the Rev. Abner Sage, Pastor of the Marked Tree Methodist Church and secretary of the planters' union, raised a howl of protest. He condemned the Editor of *The Christian Advocate* for printing an appeal by a young Emory University student who asked for clothes and funds to feed and clothe those who had fallen victims to the greed and selfishness of some of this preacher's church members.

"Naw, suh," said a widow of fifty, "we ain't had nothing to eat in four days, my little girls, four of them here, they ain't hardly got no clothes at all and they's all barefooted. I tried to make a crop last year but a widow woman what ain't got no man to help her can't hardly make no crop no how. Last year we didn't get nothing and now Dr. T—— says he's goin' to 'vict me 'cause he needs the land. We jest do the best we can, but de Lawd'll provide somehow." Great tears stood in the small blue eyes of this mother of four as she talked to me in front of her cabin. Soon a gnarled hand brushed the tears from her wrinkled and weather-beaten face. She apologized and asked me in. Once inside I saw what I have seen in a hundred sharecroppers' homes, the same misery and hopelessness pervading everything. At four that afternoon there were groceries from a nearby store and empty stomachs became full. Shortly thereafter a letter of appreciation written in pencil on cheap tablet paper came from this brave woman. Stamps cost money, and paper and pencil are not easily found among poverty-stricken sharecroppers. Three cents for a stamp is a trifle to most of us but a lot of money to a hungry and half-naked family about to be evicted from their home. In Arkansas the planters and some of the preachers call such people as this woman represents "shiftless and no count."

In the richest country on the face of the earth the braintrusters of the Wallace type, the Chamber of Commerce and the *Saturday Evening Post*, tell us that the reason people are out of work, go naked and starve, is because there is over-production. *Would to God that these eminent gentlemen were sentenced for a year to support themselves and their families by raising a crop under such impossible conditions and to feel for once the hell and misery of under-consumption.*

"Groups of white and colored workers," writes Frazier Hunt, who has already been referred to, "—men, women and children—marched across the fields, each on his own row. *In some strange way they reminded me of Chinese coolies working in the fields of soya beans along the South Manchurian Railroad. . . . Only there in distant north China there happened to be no children in the fields. Here in our own Mid-Southland no one was bothering to protest against child labor.*" [8] (Italics mine)

When a child reaches the tender age of six he actively enters the service of King Cotton. From then until he dies his life is no longer his own. His play life, his school life, his home life, in fact the whole of life, is regulated by cotton. The Children's Bureau found that nearly seventy per cent of the children of white farmers in North Carolina worked in the cotton fields and that more than seventy-five per cent of the children of Negro farmers did the same. In Texas seventy-five per cent were found to be employed in the cotton fields between six and sixteen years of age. Nearly thirty-five per cent or more than one-third of these children were under ten years of age. A child is supposed to have some time in which to romp and play and grow up. At ten and twelve the average boy in the cotton country begins to think and act like a man. At fourteen or fifteen he will marry. Life for this child has been crowded with hard and monotonous toil but when he marries and assumes family responsibilities he mires deeper into this economic morass from which he soon finds there is no possibility of escape.

"I aim," said an Arkansas sharecropper, "to give all my children a year of schoolin'." When children are needed in the fields to help the parents make a living there is not much time to go to school. A few weeks one year, a few weeks or months another year make up the child's "year of schoolin'." The schoolhouse is likely to be a long way off and the roads muddy in winter and hot and dusty in spring and fall. The schoolhouse, too, is likely to be a dilapidated one-room frame building with an ancient stove for heating purposes, rough-hewn seats from which youngsters' legs dangle in mid-air, and broken window panes. Sometimes there are holes in the floor large enough to permit a small boy's swift escape easy from an exasperated, underpaid teacher who tries to impart some notion of the three R's to her motley brood. Sometimes logs are placed against the sides of the rickety structure to keep it from toppling over and again rough lumber or tin roofing may serve to patch a broken window. There may be a consolidated school with a bus to carry the children to and from the building, but many a sharecropper's child never heard of a consolidated school and would not know a school bus from a Rolls-Royce. Thinly clad, often without shoes, trudging to school with no books, paper or pencil, to be there all day without anything to eat is not a very pleasant thing for a normal youngster.

In the cotton country a child may know the value of an education but it rarely comes within his reach. Children exist for the sake of producing cheap cotton for planters, mill owners, middlemen, merchants, and every person who enjoys paying what he calls a cheap price for a yard of cotton cloth. Cotton rules the schoolhouse as it does everything else in the cotton country. The terms of the school year are conditioned by the needs of cotton. When the child can be spared from labor in the field, the schoolhouse doors swing open; but when cotton beckons, the doors close in the face of thousands of youngsters who are an integral part of American life.

"I will be glad," said an Oxford, Mississippi, planter to me in 1928, "when the free schools are abolished. Our tenants don't need them and we can get along without them." What kind of people are these that "don't need" education? Their parents are good, kind, generous, hospitable, hard-working Americans. They love their children as few Americans today know how. I have seen mothers and fathers agonizing over the children's lack of clothes, shoes, books, pencils and paper. I have seen these sharecropper children walk ten miles to a country schoolhouse. The South is the black spot on the nation's educational map. Is it any wonder that the number of native-born illiterates in the six states of Georgia, Alabama, Arkansas, Louisiana, Mississippi and Texas is 1,286,281! As long as children are geared into the economic life of the South as they are today in the cotton fields, we need not expect to "advance one single step in civilization" but rather expect to remain that complex of fear, ignorance and poverty we are universally known to be.

Although the South is frequently referred to as "The Bible Belt," it is a land where the great masses of cotton workers are unchurched. In many areas churches have disappeared. Sharecroppers cannot afford to pay a minister and they are not welcome at the planters' church for they are frequently considered beyond the pale of those whom the church can "save." The ordinary sharecropper is not unmindful of the rôle many preachers play in giving lip service to the Master while helping "their masters" continue their exploitation and oppression of the cottonfield workers.

Each Sunday a thousand pulpiteers and Sunday School teachers thunder, "The home is the foundation of civilization and the mother's place is in the home." Let those who wear the clothes and mouth these pious phrases come to the cotton country and answer my question: "How can there be any home when mothers are forced to toil for a mere existence in the fields along with their husbands and children?" Cotton culture has meant "hands" and "hands" have meant more children and more children have meant more mouths to feed and backs to clothe and increasing poverty and degradation to the sharecropper. It is a vicious circle in which cotton pushes its slaves. In a study of the welfare of children in certain cotton-growing sections of Texas the Department of Labor found that more than 85% of the black mothers and more than 50% of the white mothers worked in the fields. The amount of rest a mother can have before and after confinement is determined by the demands of the cotton crop. I have seen a two-weeks' old baby wrapped in quilts and laid in a furrow while the mother worked the cotton. I have seen mothers ready for child-birth still in the fields pulling at the soft white fiber.

"One great reason," writes a Georgia planter,[9] "why cotton is so plentiful and cheap, is that on the great plantations of middle Georgia, middle Alabama, Mississippi and others of the richest regions of the South, it is grown by Negroes who get for their labor only enough to maintain a bare, brute subsistence. . . ." *What the Georgia planter failed to say is that not only Negroes who work in the cotton fields but whites as well, "get . . . only enough to maintain a bare, brute subsistence. . . ."* If all of the toilers who could verify this statement through actual experience were gotten together, they would constitute the majority of cotton workers in the fields of the South. The census for 1880 estimated the tenant farmer's income to be around $250. In off years it has gone much below that figure and in good years it has averaged a dollar a day and sometimes over. A study of Negro tenant farmers in Macon County, Alabama, in 1932, revealed that 61.7 per cent "broke even," 26.0 per cent went in the hole, while only 9.4 per cent made any profit at all. The average income of those who did make a profit ranged between $70 and $90 for the year.[10] The average gross income of the sharecropper in Arkansas, white and colored, in 1934, is estimated by various authorities at $210. This represents his *total* gross income for his entire year's labor and includes, of course, the labor of his entire family. Can anything more than a "bare, brute subsistence" be expected from such an income?

When the cotton has been picked and ginned "settlement" time is near at hand. To some tenant farmers it means a little ready cash for Christmas and a chance to "catch-up" on some badly needed things for the children. But to the great masses of sharecroppers "settlement day" is "the day of doom." Although a day of doom, it is looked forward to with a certain expectancy. They may get a break and come out even; they may even have a little coming to them but most of them will be "in the red." The planter "keeps books" and is "always careful that no laborer exceeds his account," and makes "sure that at the end of the year he (the planter) has gotten it all, and his labor has just lived." The sharecropper keeps no accounts, no books; it is useless to do so. When the little slip of paper is handed him bearing figures in black and red he makes no comments, asks no questions. He just looks at them, talks about them to his neighbors, goes home to his wife, dejected, downhearted, hopeless. Another year has come and gone; he has worked hard—hard as a slave—and what has he now? Nothing but a slip of paper with little figures in black and red. Another year of grinding toil faces him; maybe he will do better next year. There is hope for a while, but not for long, for he knows deep down in his heart that the future is hopeless. A great loneliness and despair seize him, and he sits down and with hands grown large from heavy work brushes away tears from eyes glued to the cotton stalks in his doorway.

After studying 1,002 farm families in Alabama, Dr. Harold Hoffsommer says, "It appears . . . that the average sharecropper cannot reasonably expect more than a bare living and that his characteristic situation is that of indebtedness to his landlord." [11]

How much did the average sharecropper in Arkansas actually make for his entire year's labor in 1933? Mr. Betts of the *St. Louis-Dispatch*, after a careful investigation, estimated his income to be $210. He arrived at his figures in the following manner. Twenty-five acres was reckoned to be the size of the average sharecropper's farm. In the rich bottom lands he would produce on an average of 1,888 pounds per acre which sold on an average for 9.4 cents per pound. The entire crop when harvested would be worth $441.90 of which the cropper's share would be $220.95. The AAA, however, forced him to plow under approximately eight acres which left only seventeen from which cotton could be harvested. This reduced his cash income to around $150. Theoretically he was to receive $60 from the Government for his share in the plow-up. This added to $150 received from the sale of the cotton which he produced amounted to $210. The record clearly shows that the majority of sharecroppers did not share in the Government benefits but that the money they were supposed to receive went to the planter. But assuming that the sharecropper did receive $60 from the Government, it would simply mean that this $210 represented the sharecropper's total income for an entire year for the labor of himself and his whole family.

Besides the "furnish" made him by the planter and which is deducted from his earnings, the planter makes other additional charges. Quoting Mr. Betts again we read: "The owner reserves the right to employ any extra labor he deems necessary and charge its cost to the cropper's account. The cropper agrees to furnish the necessary labor to keep ditches open, to pay ten cents per acre for maintaining roads and ditches on the plantation, and to pay a supervision and management fee, which custom has fixed at ten cents on the dollar of advances at the store. The latter is not a ten per cent interest charge, as none of the advances is for a period of as much as a year. Some may be for only a few weeks." [12]

Dr. William Amberson fixed the average income of sharecroppers, on the basis of a study of some three hundred families, at $262. To reach this figure, however, he estimated the rental value of the shack the cropper lived in at $50.

Sometimes the newspapers carry a brief notice of a death in the sharecropper country. It says that, "John Smith, tenant farmer, was killed today by William Adams, manager for the Jackson plantation." John Smith dared to question the figures on a slip of paper handed him by the plantation manager. For "disputing the word of a white man" he was killed. If he happens to be a white man, he may get "kicked off the place for not minding his business."

I was taking a union organizer home from jail last winter and he told me why he had no crop and had landed in jail. "I come from Tennessee," he said. "I thought I might help myself by coming down here but I ain't never seen such hard times in all my life. I always got along somehow till I come down here. I got a little education and I know something about cotton farming. The reason I've got no crop this year and hadn't none last year is because I had some sense and knowed what my right was. These planters down here, they don't want nobody with any sense working for them. They do all their business with a 'crooked pencil' and they ain't got no use for a man like me who can read and do a little figgerin' for hisself." As we rode home I thought of something I had heard as a boy in Virginia, "What the planters want is workers with brawn and not brains."

"The sharecroppers are a lazy and shiftless lot of poor whites and Negroes," declared the Rev. J. Abner Sage of the Marked Tree Methodist Church before some seven hundred people in Lottle Rock during the hearings on the Arkansas Sedition bill. Sharecroppers are no more lazy or shiftless than the system under which they live makes them. A Mississippian, Mr. H. Snyder, writing in *The North American Review,* says: "Certainly the common run of people in the South are poor, and we are told this poverty is born of their laziness. But this is upside down, as their laziness is born of their poverty." [13]

A government investigator sent to investigate the conditions of the sharecropper country was deeply impressed by the kindness, hospitality, honesty and courage of these people whom the Rev. Sage described as "lazy and shiftless." The government investigator was also impressed by the planters, riding bosses, deputies *et al.*, some of whom attend the Rev. Sage's church in Marked Tree. These friends of the Rev. Sage impressed this investigator as being more ruthless, vicious and callous to human suffering than Chicago gangsters.

Mrs. Naomi Mitchison, one of England's famous women, author of *Vienna Diary* and numerous other works of historical importance, was in Arkansas for several days in the winter of 1934. Mrs. Mitchison visited in the homes of numerous sharecroppers, ate at their humble tables and spoke at their union meetings. She writes: "I have traveled over most of Europe and part of Africa but I have never seen such terrible sights as I saw yesterday among the sharecroppers of Arkansas. Here are people of good stock, potential members of a great community, and they are being treated worse than animals, worse than farming implements and stock. They are not shiftless, they want work. They want to live decently as workers; but ever the right to work is denied them. They seem to be denied all of their rights.

"They are dressed in rags, they have barely enough food to keep them alive; their children get no education; they are a prey to diseases which the scientific resources of modern civilization could easily eliminate. I saw houses, if one can call them that, in which whole families lived in conditions of indescribable misery. Here was a log cabin half sunk in flood water and in it eight people, one of them a mother yellow and boney with malaria with a newborn child in her arms. The only furniture in the house was a table, a bench and a stove and two beds for everyone. In another home a bed this time out of old bits of rusty iron, patched with rags. The youngest child was two years old but his mother was still nursing him; she could at least be sure he got some milk that way. She herself was gray haired with a face of such misery that it seemed scarcely possible she could go on living. She earned enough to feed her children on cornbread and perhaps gravy by picking wood out of the river and selling it. Her two sons helped her do this; she could not send them to school. But even if they had not how could they have gone, wearing nothing but their rags, barefoot and full of pellagra? Everywhere the children are rotten with deficiency diseases; they are marked for life. Everywhere they are undersized; they will never grow into the strong men and women they should be.

"There is only one hopeful thing about the situation and that is the Southern Tenant Farmers' Union. Here one may see the truest human values, brotherhood and loyalty and immense courage in the face of danger and here something has happened of terrific historical importance. For the first time in the history of the United States, perhaps in the history of the world, white and colored people are working together in a common cause with complete trust and friendship. They are working together for what is supposed to be everyone's birthright —a decent standard of living, education, security, hope for the future. At present they have none of these things; their only hope of getting them is through their union. The central government has failed them, not through its own fault, but through the deliberate obstruction of the planters and their social system. It is quite clear that the planters want to keep the sharecroppers in a state of slavery. Up to now they have managed to do this. But the eyes of the world are on them. For the sake of all we value in civilization, the present state of things has got to be finished. The sharecroppers have got to be treated as decent people, the fine men and women whom they are. It will be better if this is done constitutionally and peaceably. But, if such means fail others must be used. Justice carries a scale of peace, but it also carries a sword." [14]

Weighed down by indescribable misery and suffering, exploitation and greed, the southern sharecropper has borne a mighty empire on his sturdy but weary shoulders. Honest, courageous, and long-suffering, he struggles today against the combined forces that have kept him enslaved so long. The Kingdom of King Cotton totters as the sharecropper begins to measure his strength.

IV. THE SHARECROPPER RISES

"There is only one hopeful thing about the situation and that is the Southern Tenant Farmers' Union."

—Naomi Mitchison

The Secretary of Agriculture told a Boston audience in April that the stirring of the sharecroppers and tenant farmers in Arkansas, and the Mississippi Valley generally, was due to "communistic and socialistic gentlemen" who "have gone in to stir up trouble in a sore spot." Since the "purge of the liberals" from his department—who apparently, at least some of them, understood the fundamental difficulties in the situation and were honestly striving to do something of genuine helpfulness for the sharecroppers—the Secretary has consistently ignored the real issue in what he is willing to admit is a "sore spot." What the Secretary ought to know by now is that the rising consciousness of sharecroppers in Arkansas and the South generally is due in a large measure to the stupidity and viciousness of his department policies. Had the Administration inaugurated a program which would have offered some economic security and contained some elements of justice and decency it is unquestionably true that these sharecroppers would have been the first to accept the proposals. Instead, the United States Government lent a hand in the oppression of the sharecroppers, and they became the victims of a deeper and more vicious form of exploitation than the planters alone had been able to wield. This was too much for even an humble sharecropper to understand and swallow without a protest. How the AAA functioned in the cotton country has already been referred to.

Fundamentally, however, the rise of the sharecroppers was as inevitable as the cultivation of cotton. Seventy years of struggle against a condition bordering on slavery had gotten him nowhere. The record clearly shows, moreover, that for tens of thousands of agricultural workers the situation was daily becoming worse. Seventy years had not only intensified the prevailing misery common to the masses of sharecroppers, both black and white, but had witnessed the gradual accumulation of bitterness and despair. Seventy years, it should be observed, is a long time, and even a sharecropper can learn a thing or two about life in the cotton country over such a period of time.

It is easy enough for those in high places to dismiss the grievances of the disinherited sharecroppers of the South as the trouble-making of Socialists and Communists. More than one respectable head has reposed in a guillotine basket because of misjudged historical forces and the temper it had created in men's hearts and minds. It is a well recognized fact that American workers are slow to organize, and that they do organize only when they have genuine and well-founded grievances. No man, or group of men, no matter how eloquent their tongues may be or how reasonable their ideas, can foist an organization upon American workers, least of all, upon southern agricultural workers. Americans in trade unions, and similar instruments for expressing a collective will, organize only after repeated efforts to gain anything individually have ended in failure. Similarly, it is known that workers are reluctant to join unions because of the hardships such membership frequently bring upon the workers and their families. When they do organize into a struggling union, one may admit immediately that the workers have genuine grievances. It is worth observing that men do not struggle against ancient wrongs unless they are first convinced that these wrongs are not ordained by God and are neither inevitable nor necessary to human existence, and that by struggling against these wrongs a change for the better can be effected. It would be a fool indeed who would struggle dangerously just for the sake of struggling.

We conclude, therefore, that the AAA policies of the federal government intensified the already deepening misery of the southern sharecropper and acted as a climax to the long and bitter struggle which he had been waging against poverty, disease, ignorance and semi-slavery.

Because of the furor which has been raised in Arkansas and Washington over the alleged "meddling of socialists and communists," it is well to point out at the beginning that the Southern Tenant Farmers' Union was an indigenous movement, springing up out of the very soil which bore the sharecroppers' bitter grievances. It should be said, furthermore, that the Southern Tenant Farmers' Union is not, nor ever has been, an adjunct or organ of either the Socialist Party or the Communist Party. That individual members of these parties have been prominent in the work of the Southern Tenant Farmers' Union we do not deny, but to the contrary we affirm and are proud of their achievements. Probably the reason that they were asked to assist the sharecroppers in the formation of the union was due to the fact that experienced working-class leaders were the only ones who were not on their backs, and who had the intelligence, love of human justice and courage to help them.

Just south of the little town of Tyronza, in Poinsett County, Arkansas, the Southern Tenant Farmers' Union had its beginning. In the early part of July, 1934, twenty-seven white and black men, clad in overalls, gathered in a rickety and dingy little schoolhouse called Sunnyside. The schoolhouse was old and it had witnessed many strange sights but none so strange as the one being enacted between its four leaning walls that hot summer night. Dimly lighted kerosene lamps cast strange shadows upon the faces of the men as they talked. On rough-hewn benches sat white and colored men discussing their common problems in a spirit of mutual regard and understanding.

Little time was lost in agreeing that they should form some sort of union for their mutual protection. Their main problem was to secure for themselves and their fellow-sharecroppers their share of the benefits granted under the AAA contracts. The contracts entered into between the landlords and the Secretary of Agriculture gave the tenant farmer very little and the sharecropper next to nothing but something, and they considered it worth struggling for anyway. Wholesale violations of the contracts by the planters were occurring daily. Tenants were not getting their "parity payments"; they were being made to sign papers making the landlords trustees of the bale tags; landlords were turning to day labor at starvation wages; the AAA was making things worse. They knew they could not get anything trying to dicker with the landlord individually: he would just "kick them off the place." Most of them had never been in a union and they scarcely knew how to go about organizing one. Anyway, they had to have one and they were sure that they could find someone who would know how and who would help them.

The meeting had not been in progress long when the inevitable question, which always rises when Negro and white men come together to consider joint action, arose. "Are we going to have two unions," someone asked, "one for the whites and one for the colored?" It had been many a day since the ancient walls of the schoolhouse had heard such silence. The men had thought about this before they came together, but now they had to face it and it was not an easy question to answer. A lot depended on their answer, certainly the future of their organization and maybe the future of a whole people. It was time for deep silence and exhaustive thinking. Finally the men began to speak. One man believed that since the churches divided the races that maybe the union should do likewise. He wasn't sure though, for he had noticed that the churches hadn't done very much to help the sharecroppers and that some of the church members were the cause of their suffering. Another observed that it would be dangerous for the white and colored people to mix together in their union and he was sure that the planters wouldn't stand for it. An old man with cotton-white hair overhanging an ebony face rose to his feet. He had been in unions before, he said. In his seventy years of struggle the Negro had built many unions only to have them broken up by the planters and the law. He had been a member of a black man's union at Elaine, Arkansas. He had seen the union with its membership wiped out in the bloody Elaine Massacre of 1919. "We colored people can't organize without you," he said, "and you white folks can't organize without us." Continuing he said, "Aren't we all brothers and ain't God the Father of us all? We live under the same sun, eat the same food, wear the same kind of clothing, work on the same land, raise the same crop for the same landlord who oppresses and cheats us both. For a long time now the white folks and the colored folks have been fighting each other and both of us has been getting whipped all the time. We don't have nothing against one another but we got plenty against the landlord. The same chain that holds my people holds your people too. If we're chained together on the outside we ought to stay chained together in the union. It won't do no good for us to divide because there's where the trouble has been all the time. The landlord is always betwixt us, beatin' us and starvin' us and makin' us fight each other. There ain't but one way for us to get him where he can't help himself and that's for us to get together and stay together." The old man sat down. The men decided that the union would welcome Negro and white sharecroppers, tenant farmers and day laborers alike into its fold.

When this question had been cared for the men turned toward the formation of their new organization. A white sharecropper of great ability, Alvin Nunally, was elected chairman. A Negro minister, C. H. Smith, was chosen vice-chairman. An Englishman with a ready hand for keeping minutes and writing letters was chosen secretary. A holiness preacher was elected chaplain. Some of the men had belonged to lodges and they wanted to introduce all of the secret rigamarole of the fraternal societies into the union. Some of them had formerly ridden with the Ku Klux Klan and they thought it would be a good idea for the union to operate in secret and for the sharecroppers to ride the roads at night punishing dishonest landlords and oppressive managers and riding bosses. One of the men had formerly been a member of the Farmers' Educational and Co-operative Union and advanced the idea that it would be best to have the union made a legal organization and for it to operate in the open. The men agreed that this was the best policy and they turned their minds toward getting it accomplished.

They did not know exactly how to go about having the union made legal. They could not expect any aid from any of the planter-retained lawyers and they did not have any money for fees anyway. Someone suggested that H. L. Mitchell and Clay East might help them. Mitchell was the proprietor of a small dry-cleaning establishment in Tyronza and, next door to him, Clay East ran a filling station. The two men were known throughout the countryside as "square-shooters" who always gave the underdog the benefit of the doubt, and who were, to the amusement of most people, always discussing strange ideas about labor, politics, economics and most everything else. Mitchell had once been a sharecropper himself and all of his people were farmers. He knew something about almost everything. He was known generally as a Socialist but to the landlords and planters as a "Red." Some months before Mitchell had brought Norman Thomas down to Tyronza to speak to an overflow audience at the schoolhouse. After Thomas had visited some of the plantations, had talked with the sharecroppers in the fields and at their homes, he told the audience gathered in the schoolhouse what he had seen. The planters were there *en masse* and so were the sharecroppers. The planters writhed in their seats and the mouths of the sharecroppers stood agape as Thomas denounced the system of semi-slavery under which they all lived. The planters did not forget Mitchell for this nor did the sharecroppers.

A committee from the sharecroppers called on Mitchell and East. The two men told them that while they did not know much about running a union they would be glad to help all they could. Mitchell subsequently became the secretary of the union and East its president. Mitchell wrote letters to people all over the country asking for help and advice. J. R. Butler, an ex-school teacher, sawmill hand, hill farmer, or as he prefers to be known, "just an Arkansas Hill Billy," came down to write the first constitution and to help spread the organization among the cottonfield workers. Ward H. Rodgers, a young Methodist minister with an eye to social justice, who had been preaching in the Ozark Mountains, decided to come down and preach to the sharecroppers and planters. He gave up his churches and soon became Mitchell's right-hand man.

Each night would witness Mitchell's and East's battered old automobiles loaded down with sharecroppers going to some outlying church or schoolhouse to organize a local. Enthusiasm among the sharecroppers ran high and they talked "union" with the abandon of a backwoods' revivalist. New locals sprang up everywhere the organizers went. The people were hungry for the "New Gospel of Unionism."

When the union was first formed the idea of organizing sharecroppers rather amused the planters. They generally poked fun at the boys who were scouring over the country in all kinds of weather organizing new locals. One man was heard to say, "Oh, well, let them try to organize the sharecroppers; they won't succeed for the sharecroppers are too lazy and shiftless to ever amount to anything worthwhile." But as the union continued to grow and the planters saw the Negro and white people getting together they were not quite so amused as formerly. When a car-load of white and colored sharecroppers would start off together to a union meeting someone would say, "The Socialists and the Republicans (meaning the Negroes) have joined hands." It had been rumored that Mitchell and East had "political ambitions," and that they were going to run for office in the forthcoming elections. The politicians and planters could not quite understand though why men who had "political ambitions" would bother with Negroes who had been disenfranchised and could not vote, and poor white sharecroppers who had not been disenfranchised but were too poor to pay the poll tax. Gradually the planters got the right idea. These men were not even mildly interested in the kind of politics for which Joe Robinson and Hattie Caraway had made Arkansas famous; they were interested in organizing the sharecroppers to abolish the planters' organized system of semi-slavery.

On July 26, 1934, the organization was incorporated under the laws of the state of Arkansas as the Southern Tenant Farmers' Union. The papers were taken out in White County and a certificate of incorporation received. The movement rapidly became a sort of sharecroppers' crusade and every day new applications for membership in the union were received. Men came from all over the countryside asking for organizers' papers. The union was definitely on the move.

When the first sharecroppers came together at the little schoolhouse in the early part of July, they had talked about the dangers involved in Negro and white sharecroppers organizing together. They knew trouble was coming sooner or later and they rather expected it sooner than later. Trouble was an old word with them. That was about all that they had ever known. It followed them as doggedly as their own shadow. Whatever trouble the planters might bring down upon their heads could not add much to what they had known all their lives. Trouble was as common as the air they breathed.

Scarcely a month had passed after that memorable meeting at Sunnyside when the first dark shadows of the black terror which was to later sweep over all northeastern Arkansas entered their lives. The union was holding a meeting in a Negro church near Tyronza. The chairman was explaining to the crowded audience the purpose of the union when into the church strode a plantation owner, H. F. Loan. Mr. Loan, from past experience, evidently expected to see a general exodus of the sharecroppers as he and his four riding bosses, with the inevitable guns swinging from their belts, came into the meeting. No one made a move to leave. Instead, someone asked him and his escort to have seats. The chairman continued with the meeting, explaining that the union was legal, that it operated in the open and that all they wanted was their rights which had been guaranteed to them by the Constitution of the United States and more lately by Section 7A of the Cotton Contracts. The time came for the local to be organized and Mr. Loan and his men were told that their presence was no longer desired or acceptable. Mr. Loan flushed all the way down to his boots, for he had never had anything like this happen to him before. Sharecroppers were ordering him out of *their* meeting. What insolence! However, Mr. Loan was placed at a disadvantage. There were hundreds of sharecroppers on the inside and outside of the building. They were enthusiastic too; even Mr. Loan could see that. He decided to leave peaceably. Before going he *demanded* a copy of the constitution of the union. He was quietly told that he could have one for five cents, the cost of printing.

It came through the grapevine telegraph that before dawn Mr. Loan was on his way to the state capitol at Little Rock to try to get the state officials to revoke the charter of the union. It was reported that he offered a thousand dollars to anyone who would outlaw the union.

From Poinsett County the union gradually spread to adjoining counties where the organizers met with stiff resistance on the part of the planters and officials of the law.

True to its name, Mississippi County had been one of the most difficult counties to penetrate and organize. Here they are at least consistent, in that the planters and "the law" have never allowed a prominent outside speaker for the union to address a public gathering. No section of Arkansas is more slave-ridden or barren of anything that remotely resembles what we ordinarily think of as being culture than Mississippi County. For this reason, if for no other, the sharecroppers of Mississippi County were among the first to ask for a local of the union. Four men went to Birdsong, seat of a church, a few stores and a couple of dwellings. The meeting had scarcely gotten under way when a mob of planters and deputy sheriffs led by a riding boss invaded the meeting. The organizers were bustled out of the church into their automobiles and told to "git and don't never come back here for we ain't goin' to have our niggers spoiled by no damn union." The men were escorted to the county line which is a regular part of the reception given union organizers.

Preacher Rodgers "thumbed his way" over the state spreading the gospel of social justice for the oppressed sharecroppers. Rodgers and another preacher, C. H. Smith, went to Crittenden County to organize a large local. They succeeded in holding the meeting without any trouble, but two days later Smith was caught while riding with a land salesman and dragged from the car by a group of riding bosses and deputies. He was taken to the woods and beaten and finally thrown in jail. The union officials were in a jam.

Preacher Smith was to be tried in three days. They had no one to defend him and they had no money with which to pay an attorney's fee. It meant that Preacher Smith would be sent to the county farm for six months or a year and that the work of the union would be slowed down among the sharecroppers. They racked their brains. Someone told them about a lawyer in Memphis who had a good record for justice toward the workingman. They went to see several lawyers in Memphis. One of them told Mitchell that he had lost his courage fighting in the Argonne Forest and he did not think he could recapture it in Crittenden County, Arkansas. Finally someone suggested that

C. T. Carpenter, a Marked Tree attorney, who was known far and wide for his fair-mindedness, courage and ability, might help them. Carpenter, they remembered, had a distinguished record before the higher courts of the state; he had long ago defended some strikers at the Singer Sewing Machine plant at Truman; and some twenty years ago he had fought a peonage case and at the point of a gun the plantation owner had driven him off his place. Mr. Carpenter agreed to take the case and he would not charge any fee.

When the union attorney went to Marion where Smith was lodged in jail, he was accompanied by forty or fifty sharecroppers. They flocked up the courthouse steps and into the corridors of the building. They were not noisy or sullen but they had a business look in their eyes. When "the law" saw the union attorney and this demonstration of comradeship by the sharecroppers they released Preacher Smith without a trial. Preacher Smith had been freed but the beating the riding bosses and deputies gave him injured him for life.

The effect of Preacher Smith's release from jail was electric. The union had won a victory. It could get its organizers out of jail. The officers of the union scheduled a big rally at Sunnyside Church. The sharecroppers declared a holiday, left their farm work and gathered from miles around to hear the speaking and see Preacher Smith. The church was packed and the yards jammed with men, women and children. Preacher Smith was there to tell his story. In his simple way the old man said that "the union" was his "cross" and that he was "glad to bear it for the union. For," said he, "the union is the only thing that will help my people—and all of you people, whether white or black, are my people." Vows of vengeance were made when the old man bared his broken and swollen body to the throng gathered at Sunnyside.

Although the union had won a victory it was increasingly clear to the sharecroppers that the opposition of the planters had hardly begun. The word was passed among the planters that the union was to be stopped. It did not matter particularly how it was done. Churches, where the union had been meeting, were padlocked, windows boarded over and floors removed, and schoolhouses were packed with hay. Tenants who were active in the union were given eviction notices long before the usual moving time, which is about January 1st. Others were sent threatening letters and told not to attend any more union meetings. Still others were unceremoniously driven from the plantations, flogged and beaten. Through the grapevine telegraph came the disturbing information that the planters were buying machine guns. The sharecroppers knew when and how the machine guns were to be used. At this open threat of violence the sharecroppers decided to prepare themselves for a general massacre.

They took down their old shotguns from their resting places, oiled and cleaned them. When they went to the hardware and general merchant-stores to buy cartridges, they were told by the managers that they had been ordered not to sell any tenants or sharecroppers firearms or ammunition. The order had come down from the planters. It was, nevertheless, a common sight during the latter part of the summer of 1934 to see a sharecropper walking along a dusty plantation road or a national thoroughfare with a shotgun slung in a businesslike fashion under his arm. Behind him, following closely, would be his wife and all their children. When the sharecroppers arrived at the meeting place, their guns were safely stacked in a nearby corner of the building in ready access should the alarm be given that the planters were attacking. We are somehow reminded of the Pilgrim Fathers who likewise went to their places of meeting armed with long rifles to repulse unfriendly Indians: the sharecropper with his shotgun and family brushing over a dusty road in August of 1934 carried a heavier burden than the early Fathers ever dreamed of. Thus it was that the sharecroppers sprang to the defense of the *one thing* which offered them hope and freedom and thus it was that the union survived the first onslaught of the enraged plantation aristocracy. Had they not responded as they did, it is altogether likely that the union would have been completely crushed during those first few summer months.

The union officials were aware of the grave dangers which those shotguns represented. They knew and *felt*, perhaps better than anyone else in the world, the terrific consequences of one slight error on the part of a sharecropper. A planter could commit all the crimes in the calendar and more besides, and it would not matter. He represented culture, society, money, "the law," power. He was power. His will ruled the lives of millions of helpless human beings who for seventy years had begged for mercy from him. Now they were no longer begging for mercy. They were standing on their legs demanding justice. The union officials knew the feelings that had been pent-up inside of these men; men whose lives had been shattered, who had almost abandoned hope but who were coming back—coming back like an

Arkansas tornado, full of fury. It would not take much to set off a spark that would shake the whole cotton country and drench a land in red blood. The planters, they all knew, were itching for a chance to wade into the union membership and wipe it out as they had done time and again to other unions. Give the planters a pretext and they would holler that the "supremacy of white civilization is challenged: down with the niggers." A race war would be on its way. Probably the presence of white men in nearly all Negro locals saved the Southern Tenant Farmers' Union from a bath of blood that would have made Hitler's campaign against the Jews resemble a Quaker prayer meeting. And there were women and children in nearly all the meetings. Even a riding boss hesitates to murder women and children—even if they are sharecroppers.

In order to remove every provocation the union officials asked the men to leave their guns behind. They decided to hold their meetings in the open fields under a bright southern sky. Here they were reasonably sure the planters would not attack them openly. At every meeting the officials stressed the utter necessity of proceeding legally, peaceably and democratically. The sharecroppers, almost to a man, abided by the requests of the organizers and officials.

Hundreds of sharecroppers, their wives and children came to the open meetings. Here under the open sky they heard union organizers explain the union and its aims. When such a meeting was held the planters, riding bosses, deputies, sheriff and others would gather on the fringe of the crowd. During the meeting they would amuse themselves by shooting over the heads of the crowd in an effort to intimidate the speaker and frighten his listeners. At intervals a heavy gun would roar and steel would split the air over the speaker's head. If a tree was nearby the bullets would whistle through its leaves and bits of twigs, bark and leaves would shower the crowd. When the performance had been completed a raucous laugh would emerge from "the southern gentlemen of the best families" who had gathered to terrorize the sharecroppers.

In following a strategy of non-violent resistance the union by no means accepted defeat. The officials immediately took the offensive but in a legal and democratic manner. Their first manœuver was to go into the courts on behalf of twenty-seven men who had been evicted from the five thousand acre Fairview Plantation owned by Hiram Norcross, a former Kansas City lawyer. Section 7 of the Cotton Contracts, which has already been referred to, was the basis for the suit. Mr. C. T. Carpenter was employed by the union to represent the men in the test case. At the first hearing before the Chancery Court the judge decided against the union. The case was appealed to the Supreme Court of Arkansas where the decision also went against the union.

While the sharecroppers lost in the courts, the attempt of the union to secure their rights created in the minds of the sharecroppers a wholesome attitude toward it. It was struggling and fighting for their rights. It made them feel warm and friendly toward each other and furnished them something through which they could express their collective wills.

Winter was coming. They had passed through the first few months of their struggle without any serious mishaps. The air was thick with flying steel and they saw the signs of the gathering storm. The planters had been unsuccessful in their effort to have the charter of the union revoked at Little Rock. Evictions, intimidation, beatings had not stopped the union's steady advance. The planters were to make one last effort to have the organization declared illegal through the courts.

Four union organizers, two white and two colored men, were working together in Cross County. One night they were holding a meeting in a Negro church near Parkin, seat of several recent murders of sharecroppers by plantation owners and managers, when the high sheriff of the county, accompanied by a large band of deputies, riding bosses and planters broke up the meeting and arrested the four organizers. The men were roughly handled and the union chaplain, A. B. Brookins, well past seventy, was brutally kicked in the face and stomach. Brookins, severely injured, was refused medical attention until the next day when the sheriff brought a doctor to see him. Before the union attorney, Mr. Carpenter, could reach the men, they were tried and sentenced. Bail was set at $500 when the case was appealed. The men had been arrested on the order of the prosecuting attorney, James Robertson, who claimed that planters ordered him to have the arrests made. The charges lodged against the men were two-fold: "receiving money under false pretense and disturbing labor." The union had no money with which to make bond for its men. Finally organizations in the East came to the rescue and bond was made. After remaining in jail for forty days the men were released.

The theory upon which the men had been arrested was that the union was an illegal organization; that it had never been issued a charter by the state and that therefore the organizers were receiving money under false pretenses. It was a well known fact that the union had been duly chartered under the laws of the state in White County. No one denies that the organizers were disturbing labor: it was their business to "disturb labor." It will be remembered by some, but the Rev. Mr. Sage of Marked Tree will have to get down his New Testament to read it, that it was said of Jesus that "He stirreth up the people."

The union organizers knew that this effort at strangulation would be the last straw. People were being evicted for union membership, their meetings were being broken up, the workers were being starved and terrorized and the government contracts violated in every detail by the planters. It was time to lay their grievances before Washington. Full of hope and faith five men were sent to Washington. Pennies were scraped together at every union meeting to send the Sharecroppers' Ambassadors to Washington. Outside of a promise made by AAA officials that a thorough and honest investigation of their grievances would be made the trip was uneventful. The high spot, or perhaps it would be truer to say the "tight-spot" of the trip to Washington occurred on their return journey. On a pitch black night the men lost their way in the Virginia hinterland. Setting out a-foot to discover their whereabouts they finally stumbled upon what they thought was a CCC camp. How they got inside the high wire walls they do not remember. They do remember seeing a window dimly lighted by a kerosene lamp. Up they went to the building and knocked at a door and entered a long room filled with cots upon which shapeless objects lay piled. They saw an old man huddled beside a stove in which the fire had long since spent itself. "We're lost," explained Mitchell. "Is this a CCC camp?" The old man did not reply but looked down at his feet. Tightly locked about his ankles were huge chains welded into a nearby post. The old man slowly raised his head and quietly said, "No, son, this is a convict camp." The boys did not stop to ask the old man how to get out of the prison camp. Over a high wire fence they scrambled to land full upon a blackberry thicket below. Scared and scratched they returned to their automobile and waited patiently for dawn.

When the delegation returned from Washington, the sharecroppers throughout northeastern Arkansas were all ears. What had the boys done? Whom did they see? What was President Roosevelt going to do? Was Secretary Wallace as bad as they thought he was? A thousand questions poured in upon the Ambassadors of the Sharecroppers. There was but one thing to do and that was to call a big public mass meeting where all of the delegates could report what they had found out at Washington and the sharecroppers could ask their unending questions. Word was passed around that on the 15th of January the Washington delegation would be heard on the public platform at Marked Tree.

While the delegation was away in Washington, the planters had kept things lively back in Arkansas. Evictions were increasing; the highways were strewn with ejected families; members were being threatened and whipped; false arrests were occurring almost daily; relief was falling off and people were starving and going half-naked. The hardest part of the winter was not over and madness stalked the land.

When Preacher Rodgers gave up his two little churches at Pumpkin Center in the Ozarks and came down to help in the sharecroppers' crusade for freedom and justice, he failed to bring any of his church members' money with him. He arrived on the scene broke and he remained that way thereafter. When he first came, he lived with Secretary Mitchell in Tyronza but the planters and some of the ladies of the Missionary Society in the local Methodist Church started a boycott against Mitchell and gradually succeeded in cutting his economic roots from under him. With Mitchell's business gone, it became necessary for Preacher Rodgers to find a job. He applied to the Workers' Education Bureau in Washington for a job. His application was accepted there and later honored in Little Rock. He got a job teaching in the FERA schools. It was his duty to teach illiterates how to read and write and to also attempt to give his pupils some idea of what was going on in the sovereign states of the union. His classes consisted of white and colored citizens. Evidently Rodgers was a good teacher or his pupils unusually apt, or was it just the ante-bellum attitude of the slave-owning aristocracy reasserting itself, for shortly after he began instructing the Negro sharecroppers how to read and write and *figure* the planters ordered an investigation. Led by Mr. Lynch, Superintendent of the Schools of Tyronza District, a committee of planters called while Rodgers was holding one of his classes. They found the material which the Washington office had recommended for its teachers and some material which had been handed him by the State Director in Little Rock. The planters had never seen any of the pamphlets before and while it was all as harmless as the usual Sunday morning sermon, Rodgers was henceforth condemned as a "Red" trying to overthrow the government. Teaching illiterate sharecroppers how to read and write and figure would probably destroy the kind of government to which Arkansas is accustomed. The following day Teacher Rodgers decided to call on Principal Lynch. They had a long conversation. Finally Principal Lynch informed Teacher Rodgers that if he did not leave town at once that a "committee of planters will see that you do." Rodgers remonstrated that this was a Ku Klux method. Smilingly Principal Lynch replied that it was.

The day following was the time for the Washington delegation to report. Before the sun had scarcely driven the morning clouds from its path, sharecroppers from all over northeastern Arkansas were on the roads leading to Marked Tree. They walked, rode mules, wagons, broken-down automobiles, anything to get to Marked Tree. Whole families joined the trek. It was to be a big day. *It was their day.* The men whom they had sent to Washington with their hard-earned pennies were going to talk to them. Men whom they all knew and trusted, men of their own kind, who had suffered as they had suffered, had been to Washington and were going to tell them what they had seen and heard. They had sent them; sharecroppers had sent Ambassadors to Washington to see the high and mighty. They came by the scores and by the hundreds crowding the little town that squatted in the middle of a cotton patch. They spilled over the railroad tracks, parked themselves on roofs of houses, perched themselves in trees, on the tops of trucks, box cars and wagons. A bright January sun spiced with a merry westerly wind added zest to the occasion.

The speakers were a little late arriving. The crowd wanted the speaking to commence: most of them had come a long way to hear what the men had to say—ten, fifteen and twenty miles—and they had to be home by dark. It was a long way, especially if you had to "foot it." Preacher Rodgers was going to preside at the union mass meeting. As the crowd grew restless waiting for the speakers to arrive, Chairman Rodgers decided to talk a little himself. The main body of what he had to say concerned the terrible condition facing the needy and the unemployed in Poinsett County. Relief was utterly inadequate, he said. People were starving to death. He knew. He had been in their homes; had eaten at their humble tables; had slept on their corn-shuck mattresses; had seen their hunger-ridden, starved, emaciated bodies; had heard their heart-rending and agonized cries for food and clothes and work: had felt their intolerable burden weighing down on his own head and heart. His sensitive nature made him intensely aware of the plight in which the sharecropper found himself. He suffered as they did but in a different way. He merged his intellect and emotion into the sharecropper's cause. He heard their cries, saw their anguish and spoke their mind. Preacher Rodger's voice boomed in the winter air like an angry surf at high tide. The sharecroppers crept closer and put their ears to the wind. There was a relief agency at Harrisburg, Preacher Rodgers roared, but a poor sharecropper could not get enough food there to keep the house cat alive, much less a family. When they went there, they were treated like lepers, as if they were not citizens and had no right to government relief. It was *their* money the Federal Government was spending for relief. Hadn't they made it and hadn't the planters robbed them of their wages? Hadn't they made Cotton King and clothed the world's nakedness? Now when they were cold and hungry and their children were slowly dying from starvation they went to the relief station and were treated like animals and told to "come back next week" or "you got your order last month." Preacher Rodgers said what they needed in Poinsett County was a good hunger march to demand adequate relief. Preacher Rodgers denounced the lawless acts committed against the members of the union and the attempts of the planters to destroy the sharecroppers' only hope of salvation. The Southern Tenant Farmers' Union was a legal organization, he said, chartered under the laws of the State of Arkansas; they worked in the open and only asked for justice. They hadn't committed any crimes, hadn't committed a single unlawful act yet their members had been driven from the plantations and their things dumped on the highway; their organizers had been beaten and whipped, illegally arrested and thrown into jail. The planters and some respectable citizens had threatened to lynch some of the organizers including himself. "Well," Preacher Rodgers said, "that is a game two can play. If necessary I could lead the sharecroppers to lynch every planter in Poinsett County, but I have no intention of doing so." An unutterable quality of stillness fell over the great throng. Men caught their breath and women clutched their breasts and children. The awful silence lasted only a moment and then the air was rent with the voices of thousands of sharecroppers yelling their heads off for Preacher Rodgers. The men threw their hats into the air and their arms around one another. For the first time in many a day they had heard a man say exactly what they had all wanted to say and could not. Like lightning out of the blue sky, a voice had spoken their innermost convictions and thoughts.

Rodgers was not speaking before that great multitude of the disinherited that January afternoon: it was the tens of thousands of disinherited, oppressed but nevertheless defiant sharecroppers who after seventy years of tyranny, terror and suffering were hurling their deepest feelings at the world which had oppressed them—and their voice was that of a young Methodist Preacher.

After the applause subsided and the men had recovered their worn hats, Chairman Rodgers introduced the first speaker. The first speaker for the Washington delegation was the Rev. Mr. E. B. McKinney, pillar of granite in a weary land. Chairman Rodgers as always becomes a true Southern gentleman and Christian introduced the speaker, whose brown face gave assurance to the great throng as "Mr. McKinney." On the outskirts of the crowd where the prosecuting attorney, Fred Stafford, the high sheriff of the county and all the "Better Americans" were assembled, a murmur stirred the already stiffening breeze, "He called that nigger mister."

Preacher Rodgers walked to the edge of the platform. As he jumped down the deputy prosecuting attorney, Fred Stafford, formerly of Detroit and a native of Canada, placed him under arrest. He was charged with "anarchy," "attempting to overthrow and usurp the Government of Arkansas," and "blasphemy."

V. ARKANSAS HURRICANE

"There is a reign of terror in the cotton country of eastern Arkansas. It will end either in the establishment of complete and slavish submission to the vilest exploitation in America or in bloodshed, or in both. For the sake of peace, liberty and common human decency I appeal to you who listen to my voice to bring immediate pressure upon the Federal Government to act."

—Norman Thomas over NBC

"Black terror stalks the cotton fields as the Southern Tenant Farmers' Union fights for the elemental rights of freedom of speech and assemblage—the very foundations upon which all free government rests."

—*Union Bulletin*

After a lengthy interrogation by the deputy prosecuting attorney, Preacher Rodgers was placed in jail. Shortly after he had been lodged in the small brick structure which serves as the jailhouse, sharecroppers gathered about it by the hundreds. Some wanted to tear down the structure "brick by brick and get Rodgers out." Preacher Rodgers assured his would-be-deliverers that he was much safer in jail than on the outside where he would be a ready victim for mob action by the aroused town farmers. With indignation and resentment burning in their breasts the sharecroppers followed his advice and slowly departed for their homes.

The events of the day had forecast the impending storm which was to break with unprecedented fury over the Arkansas delta country. The sharecroppers had come to Marked Tree in a holiday spirit but they were returning to their cabins sullen and rebellious. One of their best friends had been illegally arrested and thrown into jail. *He had violated no laws, made no criminal remarks; he had, moreover, only said in public what the planters and their representatives frequently did under cover of darkness.* They well understood why Preacher Rodgers had been jailed; it was because he had dared rebuke the powers that be, had condemned their system of tyranny and oppression and had defended the rights of the sharecroppers to life, liberty and the pursuit of happiness. To the plantation interests these things constituted treason, anarchy, blasphemy and sacrilege. Preacher Rodgers had merely turned the tables. Those who had been dealt insults and misery had at last begun to talk about using the same time-honored methods which the planters had used to oppress and enslave them.

It was plainly evident to the hundreds of sharecroppers who gathered in Marked Tree and witnessed the day's proceedings that the planters would tolerate no one who opposed their system. For years the plantation aristocracy had said that sharecroppers were "lazy and shiftless" and that "anything is good enough for them." They were looked upon as so much trash to be trampled underfoot or ruthlessly swept off of the plantations into the highways and swamps at the will of their masters. They were not supposed to have any rights, privileges or feelings. When a friend of the disinherited spoke his mind about conditions and used his constitutional right of freedom of speech he was jailed.

The struggle between the sharecroppers and the planters assumed the proportions of a gigantic conflict which swept everything else aside. To the sharecroppers their struggle was a holy crusade for freedom and justice. The planters looked upon the rising of the sharecroppers as the most serious threat on the horizon to their predatory system. They looked upon the Southern Tenant Farmers' Union with loathing and fear because it threatened their very existence. To crush completely this thing which had given the sharecroppers a new estimate of life and power became the consuming passion of those who lived on the backs of the sharecroppers. The struggle forced men to take sides. Few men tried to conceal their real feelings. Some were forced to do so in order to maintain their precarious economic position as merchants, traders, etc. Those who had decided opinions were outspoken in their denunciation of the opposing forces. The battle of words was at times as violent as the overt acts of the defenders of the system.

Before we proceed with the story of the terror which gripped northeastern Arkansas last winter and spring and continues in an abated manner today, it will be worthwhile to inquire into the several factors which produced it.

The program which the Southern Tenant Farmers' Union adopted was a mild, almost timid thing. Their maximum demands were a minimum upon which an organization could with decency ask for the right to live. The program of the union as contained in the official membership book is as follows:

"Our sole purpose in building the Southern Tenant Farmers' Union is to secure better living conditions by decent contracts and higher wages for farm labor and help build a world wherein there will be no poverty and insecurity for those who are willing to work.

"We demand enforcement of the Government Contracts particularly the Acreage Reduction Contracts and Section 7 of this Contract: that rental and parity payments be made direct to tenants and sharecroppers by the Federal Government. We demand that the eviction from land be stopped and that no discrimination be allowed on account of membership in an organization of the workers' own choosing.

"For free schools and buses to transport our children to and from school, with textbooks and hot lunches in school.

"For the right of all farm workers to organize without fear of terrorism and violence being done them by planters and their retainers.

"For decent wages, hours and conditions of all farm labor, with a scale to be determined by the cost of living.

"We propose a model contract providing the following: Adequate cash furnished during the farming season at a legal rate of interest, with privileges of trading where we please. Pay at prevailing wages for all improvements made on the property of the owner. Decent houses for each family, with use of certain portion of the land rent-free, for the purpose of growing food-stuffs for our families and live stock. Access to woodsland to secure fuel for our use and a wagon and team to haul it with. For the right to sell our cotton at market prices and to whom we please."

It will be readily observed that the program of the union was anything but extreme. It is significant to point out, however, that even these mild demands were regarded by the planters as confiscatory and revolutionary. By their extreme reaction to these simple demands one may gauge the extent to which the planters actually dominated the lives of their workers. Until the advent of the union, the right of the landlord to exploit and oppress the workers went unquestioned. Their will was, practically speaking, a sovereign will. They made the laws and executed them. Those who opposed their will were done away with either directly or indirectly. When the union denied the planters their right to exploit the workers and made certain demands of them on behalf of the tenants, they became possessed of a blind fury.

Had the Southern Tenant Farmers' Union accepted only white farm workers into its midst it would have been regarded with abhorrence by the planters. It would have been, however, a comparatively easy matter to crush such an organization by the time-honored method of using unorganized Negro workers to supplant the members of the union. The Southern Tenant Farmers' Union foresaw the dangers in such a program and brought the Negro and white workers together in a single union. Thus they set at nought at the outset one of the most reliable instruments of oppression that the planters had. As long as the Negro and white workers were divided the planters could keep both of them in chains. It should be added that had the union organized both Negro and white workers into separate locals the problem of breaking the union would have been difficult, yet it would have been infinitely less difficult than to crush them when they were united within the same organization. The tremendous potential power of the union of these two groups of disinherited workers was immediately recognized by the planters.

In the first place then, the mere existence of a union designed to advance the interests of the sharecroppers was sufficient to cause grave anxiety among the planters. The fact that this union was composed of Negro and white workers amazed the plantation interests and threw them into a frenzy of anger and fear.

The spokesmen of the plantation interests played up the inter-racial features of the union. All of the bugaboos of intermarriage, social equality, supremacy of the white race, etc., were given wide publicity by the plantation spokesmen and the reactionary southern press.

The plantation interests, led by the Rev. Mr. Abner Sage, tried desperately to break the union by starting a counter union along lily-white lines. A planter, Mr. Vernon Paul, said in a meeting in Memphis, that having a union of Negroes and whites was a bad thing and that such a thing might eventually come but probably not for a thousand years.

The red-baiting deputy prosecuting attorney, Fred Stafford, had taken great pains to point out that the trouble in Arkansas was due to "outside agitators." Again I repeat that the Southern Tenant Farmers' Union was an indigenous movement, spontaneously springing up in Arkansas and that at the time it was organized there was not one single non-resident of Poinsett County among them. The men who have led the Southern Tenant Farmers' Union are all southerners, born, bred and educated—if that means anything. Certainly the union has had speakers from all over the country address its meetings but so have the

Rotarians, Lions, Baptists and Methodists. They are not, of course, "outside agitators" as long as they represent the status quo. One is an "outside agitator" when he disagrees with things-as-they-are and wants some intelligence, imagination and Christian morality used in the governance of men's lives instead of the crass ignorance and injustice which everywhere abound. It might be said that for every "outside agitator" who has so much as spoken for the union, we can name one or more "outside agitators" who live in Arkansas and who exploit and oppress the sharecroppers. To mention only a few: The Rev. Mr. Sage came to Marked Tree after eleven years' residence at Southern Methodist University in Texas; the deputy prosecuting attorney, Fred Stafford, was formerly of Detroit and Canada; Mr. Hiram Norcross, defendant in the test case of Section 7 of AAA Contracts, which was brought by the union, was a former Kansas City attorney; the Singer Sewing Machine Company of New York; the Chapman-Dewey Company of Memphis, etc. Only one person from outside the deep South did any organizational work for the union and he worked for only two weeks. The cry of the "outside agitator" is a racket indulged in by those who are conscious of their sins against the poor and the oppressed, and they attempt to divert the attention of the public from the real issues to extraneous ones.

The planters, the *Memphis Commercial Appeal,* and other reactionary persons and sheets who make a profession nowadays of defending the Constitution and the flag, have been quick to exploit the fact that radicals have been prominent in the work of the union. Mr. Norman Thomas has probably done as much to reveal the true conditions under which the southern sharecroppers have been forced to live as any other living American. For three years he has hammered home to the American people facts of which they are at last beginning to take cognizance. For this he is probably the most hated and respected "Yankee" in America to southern Bourbons. Mr. Thomas, as everyone knows, is a Socialist. It was perfectly natural for Mr. Thomas, as a lover of human justice and liberty, to come to the aid of the southern sharecroppers. Mr. Thomas would be the first to deny that the Socialist Party has any designs on the Southern Tenant Farmers' Union or that it has in any way attempted to mould the policies of the organization. A member of the National Executive Council of the union told a Congressional Committee in Washington last spring that the nearest he had ever come to being a "Red" was when he voted for Roosevelt and the "New Deal." But that in the eyes of the deputy prosecuting attorney, Fred Stafford, was going too far, as Roosevelt, he says, "is too radical for me." We do not expect that this explanation will stop the squawking about "outside agitators," but for the benefit of those who wish to be informed it will not be amiss to give the facts.

Mr. C. T. Carpenter has thoughtfully observed that "the landlord and tenant system of the South seeks to avoid investigation and study." Some of the so-called "best people" of Arkansas have deeply resented the publicity given the state during the lifetime of the Southern Tenant Farmers' Union. The investigation conducted by Mrs. Mary Conner Myers was decidedly unwelcome. Newspaper men who came to Arkansas looking for the truth about conditions were regarded as so much poison. After a delegation of union men had returned from the East where they revealed the conditions under which they lived, not one of them was ever knowingly permitted to enter Arkansas again. They were ruthlessly driven from their homes—one home being practically destroyed by the withering shot from machine guns—and a price set on their heads. Several car-loads of planters and deputies waited on highway No. 64 most of a day to ambush and shoot down a Paramount News cameraman, H. L. Mitchell, and the author. The Rev. Mr. Sage told Raymond Daniel of *The New York Times* that, "it would have been better to have a few no-account, shiftless people killed at the start than to have all this fuss raised up." [1] The publicity attending the work of the union was responsible in a measure for some of the violence the planters brought down upon the heads of the union members and organizers. And it should be said that it was publicity that finally broke the back of the reign of terror last spring.

The Rev. Mr. Sage, who as a fellow-townsman recently wrote "serves his masters rather than the Master," was quick to reflect to Mr. Daniel of *The New York Times* one of the basic factors in the reign of terror. "We have had," said Mr. Sage, "a pretty serious situation here, what with the mistering of the niggers and stirring them up to think the Government was going to give them forty acres." [2] Mr. Sage put into that sentence his philosophy of life as well as that of a great majority of southerners. Negroes to him belong to a sub-race of men: they are not human; they are without the pale of civilization and should be treated as such; therefore they are "niggers." White sharecroppers, in his opinion, are no better. They are "poor white trash" to be trampled underfoot, exploited, kicked about, degraded and brutalized. The kind of Christianity represented by the Rev. Mr. Sage cannot pollute its pristine purity with poor whites and colored people. To have a kind and brotherly feeling toward poor whites and to respect the personality of Negroes by addressing them as mister is, according to this archbishop of the plantation interests, to commit the crime of crimes.

Now let us return to the story of the black terror that stalked night and day in the cotton fields of northeastern Arkansas. After repeated efforts of the planters to destroy the faith which the sharecroppers held in the union, a determined move was made to destroy the leaders and organizers. "Get the leaders" became the cry of the day and shortly thereafter every known leader of the union was a man marked for death.

Only a few days after the arrest and conviction of Ward Rodgers, H. L. Mitchell, secretary of the union, was threatened and subsequently he and his family were driven to seek protection in Memphis from the anger of the planters. In less than a week union blood had been spilt in a dozen different places.

Among the first victims of the rising terror were Lucien Koch, Director of Commonwealth College, and Robert Reed, a resident student from Texas. These men were sent by the student body of the college to assist in the organization of the union. In company with several members of the executive council of the union, Koch and Reed held a meeting in a Negro church near Gilmore, Arkansas. Suddenly and without warning the union meeting was invaded by a drunken mob of planters and deputies, and the two men were forced from the church. Speaking editorially *The Nation* says, "If anyone feels that 'terror' is too strong a word for what is going on in Arkansas let him read this paragraph from a letter from Lucien Koch."

The letter describes what happened at Gilmore. "Five men filed into the room, walked toward me, headed by the riding boss. They ordered me to 'come along.' I refused. They brandished their revolvers, dragged me from the seat, and kicked me from the room . . . two of them were violently drunk. I was hustled to a car on the road. Bob was too loyal to see me go alone . . . they poked guns into our faces and bellies, they kicked us, punched us . . . rough treatment started again. Drunk deputies stood around and allowed it to go on as everything must have been planned beforehand. Our lives weren't worth an Indian penny in the hands of those drunken, frothing madmen." [3]

A rescue party composed of sharecroppers started out to find the two men. On the road they discovered a rope with a hangman's knot already neatly tied. The rope intended for Koch and Reed had been lost by the mob and now reposes in the museum at Commonwealth College.

On the day following, a huge demonstration was staged in Marked Tree where hundreds of sharecroppers gathered to hear members of the union speak. At the conclusion of this meeting three union organizers, namely, Ward Rodgers, Lucien Koch and Atley Delaney, set out for Lepanto, seat of the illicit dope traffic of eastern Arkansas and western Tennessee, to hold another meeting. They drove into town and parked their car on a vacant lot, some forty or fifty feet from the street. Koch was sent to request permission from the mayor to hold the meeting. No sooner had the ramshackle Ford driven up in the vacant lot than hundreds of sharecroppers gathered about it. Rodgers attempted to make an announcement that the meeting would be held as soon as Koch returned, but was interrupted by the town constable, Jay May. The union organizers were jerked from the car and thrown into the Lepanto jailhouse. As Koch returned from his fruitless interview with the mayor, he inquired of the constable, whom he did not know, what had become of his companions. His simple request was sufficient to have him thrown into jail with the other men. They were charged with "obstructing the public streets, disturbing the peace of the city and barratry." To the uninitiated we should like to say that barratry is a crime of which lawyers only may be guilty, and as far as we are able to discover, it was used for the first time in America at Lepanto. The charge of barratry was subsequently dropped by the state and the city marshall, Jay May, told a *Memphis Press-Scimitar* reporter that, "They are liable to from $10 to $50 and as much as six months in jail. They'll probably get the jail sentence because we are going to put all we can on 'em." The United Press carried the following statement, "Sheriff J. B. Dubard of Marked Tree said, 'May had feared an attempt would be made to free the prisoners Saturday night. He asked for reinforcements and a guard of eighteen men, all heavily armed, patrolled the streets and alleys around the jail from late Saturday to Sunday night.' "[4]

During the time that the men were in jail there were repeated threats of lynching. The fact that the planters knew that the sharecroppers were incensed and aroused over the treatment of their organizers probably prevented a first class lynching. On the same night in which these men were arrested, Roy Burt, H. L. Mitchell and the author addressed a farmers' meeting in the northern end of the county. A mob came to the hall with the intention of breaking up the meeting and whipping the speakers, but were turned away by the indignant farmers who had gathered to hear the speaking. The union organizers were held in jail at Lepanto for three days without adequate food and heat, despite the fact that the weather was freezing. The floor was flooded with sewage water and refuse matter. "The Black Hole of Calcutta," declared a Union Bulletin, "would compare favorably with the Lepanto jailhouse."

In order to throttle the union and prevent its organizers from holding public meetings, the city fathers throughout northeastern Arkansas early passed ordinances prohibiting all public meetings. The elders of Marked Tree declared, "that an emergency has been created by speeches inimical to the welfare and calculated to disrupt the peaceful order of the citizens. Therefore, to protect the peace, health, and happiness of the citizens . . . it has been declared unlawful for any person to make or deliver a public speech, on any street, alley, park or other public place within the corporate limits of Marked Tree."

"Anyone can speak," said Mayor Fox of Marked Tree, "except the radicals. I'd give permission to 'most anybody to hold a meeting so long as they haven't been mixed up with the union and have not been listed in the Red Net-Work Book." [5]

In the early part of February, 1935, the union invited Miss Jennie Lee, a former member of the British House of Commons, to address a mass meeting of its members at Marked Tree. Accompanying Miss Lee were Mrs. Naomi Mitchison, famous English historian and novelist, and Mrs. Zita Baker, well known zoölogist of Oxford. A delegation of union members called upon Mayor Fox and requested permission to hold a public meeting in Marked Tree, at which the distinguished English guests would speak. The town constable, Art Bullard, and Mayor Fox stated that if Miss Lee spoke she would be arrested. The sharecroppers replied, particularly the women, that if "they arrest our English friends, we'll tear the jail down, brick by brick." Anxious to avoid any trouble the union officials announced that the union meeting would be held about two miles beyond the city limits.

Marked Tree became a "ghost town" as more than two thousand men, women and children flocked to the union meeting. Atop a broken-down wagon, the distinguished visitors from abroad spoke their minds about the manner in which democratic institutions, the gift of England to the world, were being scuttled and trampled underfoot by the blatant fascist demagogues of Arkansas.

A feeling of oneness prevailed the great throng that gathered at the cotton patch on the edge of Marked Tree to hear Jennie Lee. To reveal their feeling to those who would oppress them the union officers requested the entire group to assemble in marching order. Two by two the ragged and hungry army of disinherited marched, with Jennie Lee and the author leading, down the national highway, through the streets of Marked Tree, past the office of the mayor and on to the union hall. The march was orderly as the men, women and children lifted their heads in confidence and their voices in triumphant tones as they sang the union song, "We shall not be moved, Like a tree planted by the river, We shall not be moved."

It was later revealed that the planters had planned to fire upon the union marchers that February afternoon.

We do not desire, nor will space permit us, to give an account of all of the major acts of violence perpetrated upon the union in northeastern Arkansas by planters and sworn officials of the law. We shall only suggest, as has been our purpose in this chapter, the nature of various kinds of violence so as to give the reader some idea of the character of the terror which swept over the countryside.

During the middle of March Mr. Norman Thomas visited Arkansas. After a long and fruitless interview with the governor, J. Marion Futrell, Thomas began a tour of the cotton country in which the union had been most active. Thousands of sharecroppers turned out to see the man about whom they had heard so much. A union bulletin issued at the close of the tour tells the story in a graphic manner. It subsequently appeared in scores of papers throughout the country. "Norman Thomas spoke to thousands of sharecroppers, some of whom walked seventy miles to hear him. He saw conditions as they are, stark starvation stalking the most fertile land in all America; a land which is potentially the richest on earth now holding the poorest paid and most exploited workers in the world. While driving over the plantation roads visiting the people he witnessed one of the thousands of evictions that have recently occurred. He saw a white sharecropper's family, household goods and all, dumped upon the roadside. Upon inquiry it developed that the reason for the eviction of this family was that the father was a union member and that he and his wife had attempted to bring to justice a plantation rider who had violated their fourteen-year old daughter. After drugging and kidnapping the child, this prototype of the old-time slave driver raped her and after two weeks returned her to her father's home. Upon discovering what had happened to their daughter the husband and wife sought to have the fiend arrested. Instead of arresting the rapist, officials arrested the father on a trumped up charge of 'stealing two eggs.' After being released from jail, the father, who was a staunch union member, was brutally beaten. The riding boss, a relative of the plantation owner, still holds his job while the unfortunate family is thrown upon the roadside.

"Although the planting season was getting under way, this man and his family were thrown upon the roadside to starve. Without home, job or work of any sort the mother showed to Mr. Thomas the 'relief' which her husband had just received from the relief authorities at Harrisburg. The food which was to last them—a family of seven—for thirty days consisted of the following items: 8 cans of evaporated milk, 5 tins of processed beef, 1 twenty-four pound sack of flour, 1 twenty-four pound sack of meal, and three pounds of salt pork." [6]

This story is related here because it represents a type of violence inherent in the plantation system of the South and, secondly, because it represents a type of violence used again and again by the planters, their retainers, relief officials and others against those who are active in the Southern Tenant Farmers' Union.

The Union Bulletin closes by relating the now famous "Birdsong incident" which took place on the final day of Thomas' tour of the country. "As Mr. Thomas' party approached the little town of Birdsong, where he had been invited to speak by the officials of the church, it was noted that many new shiny automobiles were parked on the roadside. As some five hundred workers began to assemble in orderly fashion, thirty to forty armed and drunken planters led by a man who later turned out to be the sheriff of Mississippi County, forced their way to the front of the Negro church. As Howard Kester began with the phrase, 'Ladies and Gentlemen,' a chorus of voices broke out, 'There ain't no ladies in the audience and there ain't no gentlemen on the platform.' The gunmen surrounded the speakers and violently jerked Kester from the platform. Mr. Thomas, holding a copy of the Constitution of Arkansas in his hands, called attention to the excellent Bill of Rights contained in the document and asked by whose authority he was being prevented from holding a meeting and if his meeting was not legal. They admitted that his meeting was 'legal all right' but 'There ain't goin' to be no speakin' here. We are citizens of this county and we run it to suit ourselves. We don't need no Gawd-damn Yankee Bastard to tell us what to do with our niggers and we want you to know that this is the best Gawd-damn county on earth.' In spite of their guns, which were carelessly brandished about over the speakers, Mr. Thomas got in a few words before he was struck at from behind and finally jerked from the platform. Jack Herling, secretary of the Strikers' Emergency Relief Committee, was severely knocked on the head by a riding boss. The sheriff, sensing the desire of the mob to 'start something,' came forward and advised Mr. Thomas to leave at once as he could not give protection to him or to the innocent men, women and children who had gathered to hear him speak. He told Mr. Thomas if he didn't go there would be trouble and 'somebody might get hurt.' Mr. Thomas and his party were thrown into their waiting automobile and warned never to return. Several carloads of planters followed the car to the county line." [7]

An Associated Press correspondent from Little Rock arrived on the scene just as the meeting was being broken up. Not knowing exactly what was happening he met some of the men who followed Thomas' car from the churchyard. "My name is —— from the Associated Press," he said offering a friendly hand.

"We don't give a damn what your name is or who you are. Get the hell out of here and don't you write a line."

The whole story as everyone knows was practically scuttled by the Associated Press from its Memphis office. Freedom of the press is not always wanted by the press—particularly in Arkansas.

A type of violence frequently used by the plantation interests against the union members, and particularly against active organizers, was to cut off their access to food, clothing and other necessaries. An order would be issued to all stores in a district not to sell, or allow to be sold, any goods to certain individuals. A guard would be posted at the stores to see that the order was carried out. Likewise a guard would be posted near the home of the union member to see that no one smuggled anything to him. Thus scores of members were forced to live in the best manner possible under such circumstances. Dozens of organizers were forced to stay within their cabins because to venture out meant almost certain death. One family about which the author has personal knowledge, existed for more than two weeks on corn bread and water.

While violence of one type or another had been continuously poured out upon the membership of the union from its early beginning, it was in March 1935 that a "reign of terror" ripped into the country like a hurricane. For two and a half months violence raged throughout northeastern Arkansas and in neighboring states until it looked at times as if the union would be completely smashed. Meetings were banned and broken up; members were falsely accused, arrested and jailed, convicted on trumped up charges and thrown into prison; relief was shut off; union members were evicted from the land by the hundreds; homes were riddled with bullets from machine guns; churches were burned and schoolhouses stuffed with hay and the floors removed; highways were patrolled night and day by armed vigilantes looking for the leaders; organizers were beaten, mobbed and murdered until the entire country was terrorized. Some idea of the extent and character of the terror may be gained by citing a few instances which occurred in the space of ten days or from the 21st of March to the 1st of April. All of these incidents were reported by either the Associated Press, the United Press, the Federated Press or by special correspondents assigned to the fields, and may be checked by those who wish to do so.

March 21st.

A mob of approximately forty men, led by the manager of one of the largest plantations in Poinsett County, a town constable, a deputy sheriff, and composed of planters and riding bosses, attempted to lynch the Rev. A. B. Brookins, seventy-year-old Negro minister, chaplain of the union and a member of the National Executive Council.

After the mob had failed on four different occasions to lure Brookins from his cabin in Marked Tree, the mob turned their guns upon his home and riddled it with bullets. Brookins escaped in his night clothes while his daughter was shot through the head and his wife escaped death by lying prone upon the floor.

March 21st.

W. H. Stultz, president of the Southern Tenant Farmers' Union, found a note on his doorsteps warning him to leave Poinsett County within twenty-four hours. This note, written on a typewriter, was signed with ten X's and said, "We have decided to give you twenty-four hours to get out of Poinsett County."

On the following day Stultz was taken into the offices of the Chapman-Dewey Land Company by A. C. Spellings, Fred Bradsher and Bob Frazier on a pretext that Chief of Police Shannabery wanted to see him. While in the offices guns were laid upon a nearby barrel by the vigilantes and efforts were made by them to get Stultz to give them a pretext whereby to kill him. After being detained for three hours he was told by one of the men that he would "personally see to it that if you don't leave town that your brains are blown out and your body thrown in the St. Francis River." Stultz, the father of six small children, had been a sharecropper until driven from the land because of union activities. Night riders terrorized his family and attempted to blow up his home and in order to save them from almost certain death he and his family were moved to Memphis by the union.

March 21st.

The Rev. T. A. Allen, Negro preacher and organizer of the union, was found shot through the heart and his body weighted with chains and rocks in the waters of the Coldwater River near Hernando, Mississippi. The sheriff informed the reporter of the United Press that Allen was probably killed by enraged planters and that there would be no investigation.

March 22nd.

Mrs. Mary Green, wife of a member of the union in Mississippi County, died of fright when armed vigilantes came to her home to lynch her husband who was active in organizing the sharecroppers in that county.

March 22nd.

After threatening Clay East, former president of the Southern Tenant Farmers' Union, and Miss Mary Hillyer, of New York, with violence if they addressed a meeting of the union in Marked Tree, a mob drove the two into the office of C. T. Carpenter, the union attorney, surrounded the building and blocked all exits. After requesting protection from the mayor, East agreed to talk with the mob. He was told that if he ever returned to Poinsett County that he would be shot on sight. Mayor Fox finally interceded with the mob which allowed Mr. East and Miss Hillyer to leave town but formed an armed band to escort them out of the county.

March 23rd.

An armed band of twenty or thirty men attempted to kill C. T. Carpenter of Marked Tree, attorney for the union, at his home shortly before midnight. The leaders of the mob demanded that Carpenter give himself up but this he refused to do. With gun in hand, Carpenter prevented the mob from breaking into his home. The presence of his wife probably prevented the mob from shooting directly at Carpenter, but as they departed they poured bullets into the porch and sides of the house, breaking out the lights.

On the following night a committee from the vigilantes called upon Mr. Carpenter in his office and threatened to shoot him there if he did not sever his connections with the union. This he refused to do.

An item in the *New York Times* reads: "A band of forty-odd night riders fired upon the home of C. T. Carpenter, southern Democrat, whose father fought with General Lee in the Army of the Confederacy. The raid was a climax to a similar attack upon the homes of Negro members of the union."

March 27th.

John Allen, secretary of the union on the Twist plantation in Cross County, escaped a mob of riding bosses and deputies, who were trying to lynch him, by hiding in the swamps around the St. Francis River.

During the frantic search for Allen numerous beatings occurred. When a Negro woman refused to reveal Allen's whereabouts her ear was severed from her head by a lick from the gun of a riding boss.

March 30th.

An armed band of vigilantes mobbed a group of Negro men and women who were returning home from church near Marked Tree. Both men and women were severely beaten by pistols and flashlights and scores of children were trampled underfoot by the members of the mob.

March 30th.

A Negro church near Hitchiecoon, in which the union had been holding meetings, was burned to the ground by vigilantes.

March 30th.

Walter Moskop, a member of the trio which toured the East in behalf of the union narrowly escaped a mob which had gathered about his home to lynch him. Moskop's eleven-year-old son overheard the conversation between the vigilantes and informed his father of the mob's intention just in time for him to be smuggled out of his home by friends.

April 2nd.

The home of the Rev. E. B. McKinney, vice-president of the union, was riddled with more than two hundred and fifty bullets from machine guns by vigilantes. McKinney's family and a number of friends were inside. Two occupants of the home were severely wounded and the family given until sunrise to get out of the county. The mob was looking for H. L.

Mitchell and the author who were reported to be holding a meeting in McKinney's home at the time.

Shortly after Norman Thomas returned to New York he spoke over a coast-to-coast hook-up of the NBC. He opened his address with these words: "There is a reign of terror in the cotton country of eastern Arkansas. It will end either in the establishment of complete and slavish submission to the vilest exploitation in America or in bloodshed, or in both. . . . The plantation system involves the most stark serfdom and exploitation that is left in the Western world."

The reign of terror resulted in bloodshed but not in the slavish submission of the workers to the tyranny and exploitation of the plantation overlords. The union fought for its life as few dreamed that it could. At the end it counted its dead, its injured, its wrecked and blighted lives by the scores but it emerged from the struggle a powerful, well organized and fighting union which, if it continues along the road it has thus far traversed, may have a significant influence upon the future of American history.

During these days of terror the men who directed the activities of the union lived in the expectation of immediate death. Without sleep, without food and often without hope—but never without faith in the ultimate triumph of their cause—they directed a brilliant struggle along the lines of non-violent resistance which won for them the support of the entire country. Had it not been, said a *New York Times* reporter who visited the scene, for the policy of passive resistance which was adopted by the leaders of the union, Arkansas would have been visited by a terrible massacre and bathed from one end to the other in a bath of red blood.[8]

It was no easy task for men who have been reared with a shotgun in their hands and who have been taught "direct action" from the cradle up as the best method of settling disputes, to submit to the fury of the Arkansas hurricane let loose by the plantation masters against their ancient serfs.

The back of the terror was broken not by resorting to the violent tactics of the defenders of the system but by the written word and the ceaseless activity of the thousands of disinherited men, women and children who lived for the union, but were ready to die that the union might live and bring some measure of peace, freedom and security to their children.

VI. THE DISINHERITED FACE THE FUTURE

"The test of any civilization is its view of man." [1]

—Stanley Jones

At the height of the reign of terror in Arkansas, there was being exhibited to the startled eyes of hundreds of cotton producers and plain citizens of Memphis a machine which carried within itself more terror to the already disinherited sharecroppers and day-laborers than anything that has come into their lives within the memory of living man.

The cotton picking machine will come to the cotton fields of the South and when it comes it will produce a major social revolution. It will come and there is no way to stop it; it is the way of the machine and the way of America. The question all America must face is: what can be done to rescue from utter despair the millions of men, women and children whose homes, jobs and way of life will be totally altered by the coming of this machine and the increasing mechanization of cotton. If we are to avert a major social catastrophe this question must be answered quickly and decisively. Let us see what is involved.

For many years men have been trying to invent a machine that would pick cotton. Millions of dollars have been spent by various corporations and groups. Within the last few years a cotton picking machine has been produced by the Rust Brothers. The machine, we are told by technicians and producers, is about as nearly perfect as a cotton picking machine can be. Perfection, in this instance, means that the machine actually picks in one hour what four ordinary pickers can gather in a day. In seven and one-half hours the machine will do the work which it would require a good hand picker about three and one-half months to pick.

The cotton picking machine represents but one phase of the increasing mechanization of cotton—but a highly important one. Cotton, like man, has moved West until today most of the American cotton crop is raised west of the Mississippi River. In the state of Mississippi and the states west of the river nearly twice as much cotton is produced as in all of the other cotton raising states of the Old South combined.

The cultivation of cotton in the Old South has been done largely through the use of a single plow and the ancient mule. In this region the soil has been mined out by successive crops of cotton and by devastating rains which have eroded large areas. It is now almost impossible to grow sufficient cotton of a good quality on much of this land without the profligate use of expensive fertilizers.

The Southwest on the other hand is mostly new soil of a rich nature stretching for miles over flat surfaces. Here the land is well adapted to the use of tractors and large scale farming. A mechanized farm in the Southwest can produce cotton from one-half to three-quarters cheaper than a farm employing the old methods. It is inevitable, therefore, that there will be an increased use of the tractor and other mechanical equipment in the production of cotton, and it seems likely that in the near future practically all of the American crop will be produced west of the Mississippi River.

The coming of the cotton picking machine plus further mechanization in the cultivation of cotton will have tremendous social consequences in the South and in America. In the first place, large areas in the Old South which have been given over to the raising of cotton for generations will be forced to quit the business because of the impossibility of competing with mechanized production in states like Mississippi, Arkansas, Texas and Oklahoma. Large numbers of small independent farmers who have been able to make a little cash money from cotton will be forced out of the race also. Increased mechanization of cotton will completely liquidate large numbers of tenant farmers, sharecroppers, and day laborers not only in the Old South, but in those regions to which the machine is well adapted.

In analyzing the problem two other important factors must be taken into consideration. Cotton has recently met new competitors in the form of rayon and other synthetic materials. While the industry is young it is making rapid gains and bids fair to make even greater inroads upon the cotton market.

The American cotton growers face a steadily declining world market. Cotton is being produced by foreign countries on an unprecedented scale. Brazil, India, China, Russia, Australia, Syria, the Sudan, and Rhodesia are but a few of the fifty nations now producing cotton. Some of the newer cotton producing countries are growing a staple of better quality than that grown in America.

Already thousands of cotton field workers have been driven from the land. What will happen to them and to the thousands to follow as the above mentioned factors begin to lay an earnest siege to the beleaguered Kingdom of King Cotton?

When we view as a whole the havoc that will be wrought by the continuance of the economy of scarcity, by the increased mechanization of cotton, by the drying up of foreign markets, by the increased competition within the domestic market of rayon and other cotton substitutes, we may well be alarmed about the future.

The evils of tenantry have been tackled with enthusiasm and skill in nearly all civilized countries except America. In Ireland, Denmark, and other countries genuine and forthright measures have long since been taken to abolish the evils of the system, but here in America we are consumed between stupid palliatives and a death-like inertia. The solutions to the problems involved in cotton tenancy are not readily discovered nor, when discovered, easily followed. Evils that have been in the making for generations and have their roots planted deep in our national life may not be solved in a day, but the will and determination to rid our land of the evils of the system must be reached today. To postpone definite steps toward a healthy and genuine solution of the problem is but to invite disaster.

It seems to me that men today, in America and everywhere, are faced with only two alternatives. The choice lies between an economy of scarcity and an economy of abundance. The former will reward us with increased suffering for the masses of our people, greater poverty and insecurity, with intensified nationalism, increased restriction of trade and commerce, increased demand for a mightier army and navy, new world wars and a future filled with unimaginable suffering and misery.

The only sane road for men to follow is that one which leads toward an economy of abundance. Such an economy promises ultimately a full and abundant life for all. The establishment of an abundant life has long been a dream, but in our day the resources of modern civilization make it possible of realization when men set their hearts to it. Only men of great courage, social consciousness, world kinship and prophetic imagination will dare travel this road and seize the abundant life. The timid and unimaginative will want to live and die by their fathers.

Few people more sincerely long for a full life than the millions of disinherited sharecroppers and tenant farmers in the South. Yesterday they were without hope but today they have their ears to the wind and their eyes fixed on the far horizon that looks toward the fulfillment of their reborn hopes and aspirations. Yesterday they were pleading for mercy; today they are demanding justice. Yesterday they did not see the Kingdom of the Good Life; today they behold it, tomorrow they will rise up and seize it.

In an effort to ease the burdens of the disinherited sharecroppers, tenant farmers, and farm day-laborers in the South, and to mitigate some of the evils flowing from the collapse of King Cotton's Kingdom, various measures have been inaugurated by the Roosevelt administration.

The measures which have been hailed as the solution of the evils of tenancy in the South are those embodied in the Bankhead-Jones Bill and supplemented by the Resettlement Administration. In order that the reader may gain some idea of the purpose and contemplated program of the Resettlement Administration, we quote below a letter recently received from Mr. John Carter, Director of Information, Resettlement Administration.

In answer to your request for a written statement regarding the purpose and contemplated program of the Resettlement Administration, I submit the following:

The Resettlement Administration was established by Executive Order.

1. To administer approved projects involving the resettlement of destitute and low income families from rural and urban areas including the establishment, maintenance and operation, in such connection, of communities in rural and suburban areas.
2. To initiate and administer a program of approved projects with respect to soil erosion, stream pollution, sea coast erosion, reforestation and flood control.
3. To make loans as authorized to finance in whole or in part the purchase of farm land and necessary equipment by farmers, farm tenants, croppers, and farm laborers.

The Resettlement Administration has taken over the work of the following agencies.

1. The Rural Rehabilitation Division of FERA.
2. The Subsistence Homestead Unit of the Department of the Interior.
3. The Land Policy Section of the AAA Program Planning Division.
4. The Farm Debt Adjustment Unit of the Farm Credit Administration.

For the purpose of administering this program, the Resettlement Administration has been divided into four divisions.

1. The Land Utilization Division which plans the purchase of land in submarginal areas for the purpose of retiring such land from agriculture and devoting it to more valuable uses such as forestry, grazing, and recreation.
2. The Rural Resettlement Division which plans, first, to make loans or grants to some 525,000 destitute or low income farm families through its rehabilitation program. These loans and grants will be made to farm tenants, sharecroppers and farm operators. Second, to resettle some 10,000 families in rural resettlement communities which will be of three types: (1) purely agricultural communities; (2) agricultural and industrial communities; (3) co-operative communities. Third, to alleviate the distress in rural areas caused by foreclosures by making available to farmers the services of the Farm Debt Adjustment Committees who will attempt to bring about the adjustment of debts between creditor and debtor.
3. The Suburban Resettlement Division plans to furnish housing in four suburban areas for about 5,000 low income urban families.
4. The Management Division was established to direct the activities of the communities set up by the other division until these communities are liquidated.

The general program of the Resettlement Administration contemplates the expenditure of about $330,000,000, representing efforts in every state of the union through June 30, 1936.

One may be grateful that something is being attempted to heal the naked wounds of our people. However, in view of the immensity of the problem confronting us in regard to the immediate and wholesale removal of tenants from the land, not to mention the multitude of vicious circumstances surrounding the workers generally, it is plainly evident that the program of the government is utterly inadequate.

In our opinion the Resettlement Administration has not yet come to grips with the reality in the situation and its program will probably fall immeasurably short of the goal some of its most enthusiastic supporters have set for it.

The program is purely palliative and does not strike at the roots of the problem. The program strikes us as being something more of a salve rubbing orgy than an attempt to perform a major operation. It is not amiss to inquire as to the wisdom of rubbing salve on a cancer that is admittedly gnawing at the very vitals of the entire body. Tenancy as it operates in the South today is a cancerous growth and nothing short of a major operation will suffice to rid us of its evils.

The days of salve rubbing, it might be observed, have passed. The present conditions demand drastic and fundamental alteration. For the purpose of genuine reconstruction and development we need billions not millions. We need to tear down and build up not on a petty scale but on a vast regional and national one. We should be done with palliatives which paint the rotten apples red and leave them to rot everything that comes into contact with them. We need to be done with tinkering and lay hold of a definite plan that will bring justice and security to all the people.

Since I have dealt rather summarily with the proposed program of the government what have I to offer? I do not pretend to know what is in the mind of the officers of the Resettlement Administration with reference to the future. I venture to suggest, however, a line that can and must be followed if we are to move forward in the direction which we shall ultimately be forced to follow.

Land and natural resources are the common heritage of all the people of the United States. Today the land and the natural resources have become the property of a few and millions of our fellows live in want of the things which by right should be theirs and their children's. All titles to land should be vested in the citizens of the United States. The use and the occupancy of land should be the sole title to land. Without the adoption of such a policy we cannot begin to tackle the problem of tenancy in America.

In order to achieve this a National Land Authority should be established at the earliest possible moment. The Authority should be an instrumentality of the United States with all the powers thereof. It should have the power of eminent domain with the right to purchase land at a fixed sum arrived at by land experts of the Authority. It should plan the entire agricultural program of the nation and mobilize the necessary forces for its execution.

Since this volume is primarily concerned with tenancy and since tenancy is peculiarly a southern institution, what I have to say pertains largely to the South.

Such an authority would set as its task the abolition of tenancy in all of its forms. It would undertake to revamp the whole rural structure of southern agriculture. This would mean that in certain localities the Authority would eventually look toward the removal of whole populations from areas where only a subsistence and bare living can now be gotten from the soil. It would mean that the Authority would undertake to build in the rural South a high and altogether new type of rural culture. Large areas in the South which are now under cultivation yield but a miserable living for farmers and produce a backward citizenry. There are other large areas which are not being cultivated but which could be made to yield abundantly. The Authority would plan to retire the mined out and eroded lands and settle the people from submarginal lands to those productive areas. Large areas that are now being cultivated would have to be retired, reforested and put to other uses. Such a plan is part of the Resettlement Administration's program.

The Authority would set for itself the task of teaching the South the values to be obtained from diversifying agriculture. Diversification must be pursued relentlessly and scientifically.

In my opinion the effort to establish small independent farmers on land which they are to buy over a long pediod of years will get us nowhere except to establish a peasant class of farmers whose conditions will not be any better at the end of the experiment than at the beginning. My experience with sharecroppers and tenant farmers in Arkansas and elsewhere leads me to believe that co-operative farming on a large scale would be most acceptable. There are certain invaluable things which may be achieved in co-operative farming which cannot be achieved among widely separated individual farm homes. Such a plan means, for example, that the farm workers would live on a large plantation or farm. The farmers would be entitled to the land as long as they occupied and used it, and it could not be taken away from them on any account as long as they fulfilled their obligations to the community or to the state. They would work the land co-operatively, sharing the proceeds alike at the end of the year. They would own their homes, etc. but would live a group life, or a community life. By living in compact communities all of the resources of our modern world could be put at their disposal to develop a high type of rural life. All of the conveniences now afforded urban dwellers could readily be established in such a community. Agricultural engineers would bring to the people the best possible methods of getting the most from the soil and social engineers would direct them in the achieving of a high type of social relationship. A really adequate education for the children might replace the present miserable travesty of learning.

Such associations would not be established by compulsion but by example and education. If the rural workers were to have demonstrated to them the type of life that awaits them, they may be attracted away from their old ways with a minimum of social disturbance.

There are areas where large scale co-operative farming, such as I have briefly suggested, is out of the question in the immediate future, and where small aggregations or individual homesteads would be necessary and desirable from a social point of view. In such instances the Authority would render all assistance and aid at its disposal in helping the individuals or individual to accommodate as readily as possible to the changes taking place.

It will be argued that the American farmer, and particularly the southern farmer, is a super-individualist with a great longing for the land. Certain efforts have been made by the Southern Tenant Farmers' Union to discover what the sharecroppers and tenants actually think. The plan so briefly suggested above is partly determined by what we discovered to be their wishes.

To thoroughly explain all that we have in mind would require another volume and we have suggested in barest outline what course we believe should be followed. We believe that only as we push forward toward a co-operative society based on the production for use and not for profit will we solve the problems of our day.

In the meanwhile thousands of tenant farmers, sharecroppers and day laborers will continue to be the unwilling victims of the feudalistic semislave system of tenancy or sharecropping. No immediate hope can be expected from the federal government either through the administration of the AAA contracts or through the Resettlement Administration. While it is understood that the new cotton contracts will be more favorable to the real producers, who are the tenant farmers, sharecroppers and day laborers, in all probability the contracts will remain a "landlord's code" and they will find ways to circumvent and annul the theoretical guarantees to the workers. Until Washington becomes really concerned about a genuine *New Deal* for the "Forgotten Man" and gives him a voice in the administration of the contracts there will be little change in the status of the workers in the cotton country.

Men do not freely relinquish power and control over the lives of other men. Particularly is this true in large sections of the South where landlords and planters completely dominate the lives of the people. Whatever rights sharecroppers and tenant farmers achieve they will win by their own efforts. They will enter the Good Life when they are in a position to wrest it from the greedy hands which now withhold it from them.

For years workers in the cotton fields of the South have been able to secure only a "bare, brute subsistence" from their overlords. Today through the Southern Tenant Farmers' Union these once impotent men are able to secure rights which they were never able to secure alone.

On a never-to-be-forgotten day I visited some friends who were working in the radish fields of Arkansas. I saw men, women, and children bending over the earth pulling the tiny delicacies from the ground. Burned by the hot sun they worked at their tasks until the lengthening shadows cooled their aching bodies and gave them rest from their grinding toil.

For every one thousand eight hundred radishes they had pulled and tied into bunches of eighteen each, they received twelve and a half cents. As they trudged home to their little cabins some carried twelve cents, some a quarter, a few a little more as the reward for their day's work in the fields.

That night I sat down in a restaurant near the Peabody Hotel in Memphis. The waitress brought a steak. Lying beside it was a radish—a radish as red as blood. As I looked at it, I seemed to see the workers I had talked with and watched as they worked in the fields a few hours before. In the little red radish I saw the agony, the despair, the deepening terror of starvation that clutched at the hearts of millions of disinherited men, women and children throughout the South. All the misery and hell of slavery seemed bound up in that little red radish. I tried to eat but I could not, for the radish on the platter had become human blood wrung from the sweating backs and aching bodies of the disinherited.

A few months later these lands had blossomed into a white loveliness: cotton picking time was at hand. Men were asking, "How much will we get for picking?" The planters met and a tall man who might have been mistaken for a college dignitary said, "There are enough starving people in Arkansas to pick all of our cotton for twenty-five cents a hundred. We don't have to pay more." Some planters agreed to pay twenty-five cents and some said that they would pay forty cents.

For months the Southern Tenant Farmers' Union had been preparing to call the cotton pickers out of the fields on strike against these starvation wages. Many of the people were totally ignorant of what a strike was for but they readily agreed to follow the union leaders. The union set up strike committees throughout the countryside and distributed thousands of pieces of literature. Union organizers secretly visited cabins on the plantations where they met the local leaders and helped them work out their strike plans.

The members of the union knew that there would be trouble and were afraid that the Arkansas hurricane would descend once more to strike terror in their midst. They made up their minds that whatever came their way they would be ready and they girded themselves for the struggle.

The announcement that the union would call a strike brought a slight smile to the face of many planters, jeers and ridicule from some, and threats of violence from others. Certain relief authorities announced through the papers that there would be no labor shortage in the cotton fields. There were starving men on the relief rolls, they said, who would take the jobs of the cotton pickers if they went out on strike. The officials said that they would furnish trucks to transport the workers to and from the fields.

Union officials were quick to point out to the small independent merchants that if relief workers from the outside were allowed to come in to the cotton fields and take the jobs of the strikers that they would lose the money which the local workers would ordinarily spend with them. The merchants were quick to grasp what the strike meant to them in terms of more money in the pockets of sharecroppers and day laborers. They began calling relief officials over long distance. A day or so later an announcement appeared in the papers stating that no men on relief would be sent into the cotton fields.

On a given night the strike bills were distributed throughout the territory. So effectively were they distributed that many people thought an airplane had dropped them down on the cabins. On the following day a strange emptiness hung over the cotton fields. Most of the workers did not go to the fields and those who did soon returned to their cabins. Violence flared here and there as planters assisted by deputy sheriffs arrested strikers and organizers. The strikers had been told by the union not to resist arrest but to be willing to go to jail. "Fill every jail in Arkansas but don't pick cotton until union prices are met," was the message of the organizers to the strikers. Scores of men were arrested and some jails were crammed with strikers. Violence reared its head but the cotton hung in its boll as the strikers stayed in their cabins.

Six days after the strike was called the planters pushed their prices to a low of sixty-five cents per hundred pounds and a high of one dollar. The union had asked for a dollar a hundred. A labor official from Little Rock made a trip through the cotton country to see how effective the strike was. He reported that he saw two workers picking cotton. The strike was effective and the cotton hung in the bolls until the union told the men to go back to work.

In thousands of cabins in the cotton country men received a new light. They were no longer powerless: they could raise their wages and better their own working conditions when they were willing and had a mind to do it.

Of themselves these disinherited men had done what no one else could do for them. They had done what Washington with all its power and glory had been unwilling to do. They had raised themselves against the will of their masters and had won without the loss of a single man.

In the little schoolhouse at Sunnyside seventeen ragged and disinherited sharecroppers brought forth the Southern Tenant Farmers' Union. The seventeen men of faith created a movement that harbors the faith, the hope, and the collective will of twenty-five thousand black men, white men, Indians and Mexicans in six southern states of the mid-South and Southwest.

They have grown in numbers, they have grown in ideas and they have grown in power. They know that poverty and misery and ignorance and all the hellish evils of King Cotton's Kingdom are not ordained of God. They know that they can change things when the slaves of King Cotton set their hearts against the tyranny and terror of the cotton country. They know that they must build a mighty union that will sweep into hell the system that holds them in bondage. They know that the Galilean carpenter talked about the Kingdom of God on earth and they know that they alone can make his vision of brotherhood and justice a reality in this world.

These disinherited men have their ears to the wind, their eyes fixed on the far horizons where freedom and plenty await them. Today they march with firm feet toward it; tomorrow with firm hands they will seize it.

To the disinherited belongs the future.

NOTES

Chapter II

1 Johnson, Embree and Alexander, *The Collapse of Cotton Tenancy*, The University of North Carolina Press, 1935.

2 *Memphis Commercial Appeal*, Nov. 17, 1935.

3 W. R. Henry, *Cotton and the Commission Merchants.*

4 Daniel J. Sully, "King Cotton's Impoverished Retinue," *Cosmopolitan*, Feb., 1909.

5 E. V. Wilcox, "The Great White Way of Cotton," *The Country Gentleman*, Mar. 31, 1923.

6 *Ibid.*

7 T. N. Jones, *The Dallas Morning News*, Feb. 17, 1928.

8 Fred C. Kelly, *Today*, March 30, 1935.

9 The Rev. Alfred Loring-Clark as quoted in the *Memphis Press-Scimitar*, Sept. 28, 1935.

10 Daniel J. Sully, "King Cotton's Impoverished Retinue," *Cosmopolitan*, Feb., 1909.

11 John A. Todd, *The World's Cotton Crops.*

12 Harold Hoffsommer, "The AAA and the Cropper," *Social Forces*, May, 1935.

13 *The Nation*, Feb. 13, 1935.

14 Cotton Contract, USDA, AAA, Washington, D. C.

15 *The Nation*, Feb. 13, 1935.

16 *Ibid.*

17 *Ibid.*

18 *Ibid.*

19 *Ibid.*

20 Commodity Information Series, Cotton Leaflet No. 1. USDA, AAA, Washington, D. C.

21 *The New York Times*, April 17, 1935.

22 Frazier Hunt in an article written for the Newspaper Enterprise Association and published in the *New York World-Telegram*, July 30, 1935.

Chapter III

1 Norman Thomas, *Human Exploitation*, Frederick A. Stokes Co., New York, 1934.
2 Johnson, Embree and Alexander, *The Collapse of Cotton Tenancy*, p. 22. The University of North Carolina Press, 1935.
3 *Atlanta Constitution*, Sept. 27, 1914.
4 Frazier Hunt, Newspaper Enterprise Association, *New York World-Telegram*, July 30, 1935.
5 Oliver Carlson, "The Revolution in Cotton," *The American Mercury*, Feb., 1935.
6 B. J. Bivens, *The Farmers Political Economy.*
7 Fred C. Kelly, *Today*, March 30, 1935.
8 Frazier Hunt, Newspaper Enterprise Association, *New York World-Telegram*, July 30, 1935.
9 C. D. Rivers, *The Empire of Cotton.*
10 Johnson's, *Shadows of the Plantation*, University of North Carolina Press, 1934.
11 Harold Hoffsommer, "The AAA and the Cropper," *Social Forces*, May, 1935.
12 Curtis Betts, *St. Louis Post Dispatch*, March 11, 1934.
13 H. Snyder, *North American Review.*
14 Statement issued by Naomi Mitchison and given to author.

Chapter V

1 Raymond Daniel, *The New York Times*, April 18, 1935.
2 *Ibid.*
3 *The Nation*, March 13, 1935.
4 *Memphis Press-Scimitar*, Feb. 4, 1935.
5 *The New York Times*, April 21, 1935.
6 STFU Bulletin, "Norman Thomas Visits the Cotton Fields."
7 *Ibid.*
8 *The New York Times*, April 18, 1935.

Chapter VI

1 From *Christ's Alternative to Communism* by E. Stanley Jones, Copyrighted 1935, by permission The Abingdon Press.
2 Letter to author from Mr. John Carter, Director of Information, Resettlement Administration, Washington, D. C.